Nanomaterials for Direct Alcohol Fuel Cell

Nanomaterials for Direct Alcohol Fuel Cell

edited by

Yixuan Wang

Pan Stanford Publishing

Published by

Pan Stanford Publishing Pte. Ltd.
Penthouse Level, Suntec Tower 3
8 Temasek Boulevard
Singapore 038988

Email: editorial@panstanford.com
Web: www.panstanford.com

British Library Cataloguing-in-Publication Data
A catalogue record for this book is available from the British Library.

Nanomaterials for Direct Alcohol Fuel Cell

ISBN 978-981-4669-00-9 (Hardcover)
ISBN 978-981-4669-01-6 (eBook)

Printed in the USA

Contents

Preface

Direct alcohol fuel cells (DAFCs) are very promising advanced power systems that may considerably reduce dependence on fossil fuels and are, therefore, attracting increased attention worldwide. Nanostructured materials can significantly improve the performance of the cathodes, anodes, as well as electrolytes of DAFCs. This book focuses on the most recent advances in the science and technology of nanostructured materials for DAFCs, including novel non-noble or low noble metal catalysts deposited on the graphene layer and metal-free doped carbon black for oxygen electroreduction reaction, Sn-based bimetallic and trimetallic nanoparticles for alcohol electro-oxidation reaction, and novel nanomaterials for promoting proton transfer in electrolytes. In addition, the book includes chapters from not only experimentalists but also computational chemists who have worked in the development of advanced power systems for decades. The research-and-development outlook for new DAFCs materials has also been suggested to facilitate innovation in this important area. Illustrated throughout with excellent figures, this multidisciplinary work is not just a reference for researchers in chemistry and materials science, but a handy textbook for advanced undergraduate- and graduate-level students in nanoscience- and nanotechnology-related courses, especially those with an interest in developing novel materials for advanced power systems.

In Chapter 1, Drs. Youjun Fan, Junming Zhang, Qingyu Li, and Jiujun Zhang focus on the recent advancement in the anode catalysis of DAFCs, including Pt/Pd-based catalysts and other non-noble metal catalysts. They also discuss in detail the supporting materials such as carbon nanotubes (CNTs), carbon nanofibers (CNFs), mesoporous carbon, and graphene, as well as non-carbonaceous conducting polymers and hybrid supports materials. In Chapter 2,

Dr. Jayati Datta comprehensively discusses the state-of-the-art catalyst technology for anodic oxidation of alcohols, in particular for low-temperature fuel cell application, including the catalysts designed with a multimetallic framework. A great deal of interest has arisen for DAFCs working in an alkaline environment, which allows the use of inexpensive non-platinum metal as well as transitional metals metal oxides as critical electrode component harnessing more energy and at the same time ensuring affordability of the fuel cell system.

In Chapter 3, Drs. Yixuan Wang and Zhenfeng Xu summarize the systematic density functional theory studies on the adsorption and decomposition of water and ethanol on Pt_nM ($n = 2, 3, 6$, and 9; M = Pt, Sn, Ru, Rh, Pd, Cu, and Re). The results show that from both kinetic and thermodynamic viewpoints, Sn is more active to water decomposition than pure Pt and all the other investigated PtM except PtRe, which well supports the assumption of the bifunctional mechanism that an Sn site accelerates the dissociation of H_2O. Chapter 4, a theoretical work by Takayoshi Ishimoto and Michihisa Koyama, is devoted to the mechanistic aspects of ethanol and glucose oxidation in an alkaline environment to provide insights into the direct use of biofuels in fuel cells. Au is selected as a promising catalyst alternative to Pt in an alkaline environment. In Chapter 5, Liuming Yan and Baohua Yue contribute to the proton, alcohol, and water transport mechanism, and the rational design of proton electrolyte membranes (PEMs). They thoroughly discuss a guideline to design more efficient PEMs at a molecule level.

In Chapter 6, Drs. R. K. Singh, F. G. S. Wasim, and M. Neergat present the new nanomaterials, both precious and non-precious metal-based, relevant to oxygen reduction reaction (ORR). Methods for estimating the ORR product selectivity (H_2O/H_2O_2) and the electrochemical surface area (ESA) are introduced. The evolution and synthesis of the catalysts for the ORR are then discussed in detail. Overall, the investigations on ORR significantly contribute to a better understanding of the electrochemistry of nanomaterials.

In Chapter 7, I primarily summarize the effects on the activity of the electrocatalyst in DAFCs, including the composition (Pt:Sn ratio), degree of alloying, roughness and size of particles, and the pH as well as characteristics of the electrolytes of DAFCs.

I greatly acknowledge the dedication of all the authors mentioned above. Without their effort, the project could not have been finished. My appreciation also goes to Pan Stanford Publishing.

Yixuan Wang
Albany, Georgia

Chapter 1

Advanced Anode Catalysts for Direct Alcohol Fuel Cells

Youjun Fan,[a] Junming Zhang,[a] Qingyu Li,[a] and Jiujun Zhang[b]

[a]*Guangxi Key Laboratory of Low Carbon Energy Materials, College of Chemistry and Pharmaceutical Sciences, Guangxi Normal University, Guilin 541004, China*
[b]*Energy, Mining & Environment, National Research Council Canada, Vancouver, BC V6T 1W5, Canada*
youjunfan@mailbox.gxnu.edu.cn, jiujun.zhang@nrc-cnrc.gc.ca

Direct alcohol fuel cells (DAFCs) need their anode electrocatalysts to be improved in order to reach performance levels sufficient for practical applications of electric vehicles and portable electronic devices. This chapter focuses on the recent advancement in the anode catalysis of DAFCs. First, new developments in Pt-based catalysts, Pd-based catalysts and other non-noble metal catalysts are reviewed. Then the support materials aiming to enhance the activity and durability of the catalysts for the alcohol oxidation reactions are discussed in detail. Particularly, the performance issues associated with a variety of carbon supporting materials such as carbon nanotubes (CNTs), carbon nanofibers (CNFs), mesoporous carbon and graphene, as well as non-carbonaceous-based materials such as conducting polymers and hybrid supports materials are analyzed.

Nanomaterials for Direct Alcohol Fuel Cell
Edited by Yixuan Wang

ISBN 978-981-4669-00-9 (Hardcover), 978-981-4669-01-6 (eBook)
www.panstanford.com

It is emphasized that open-surface structures of nanocrystals with high surface energy and high density of low-coordinated step and kink atoms are crucial to their catalysis applications. Considering the importance of material structures, this chapter also summarizes the recent progress on the surface shape-controlled synthesis of metallic nanostructures and their crucial role for alcohol electro-oxidation. Finally, several research and development (R&D) directions for new DAFCs anode catalysts are also suggested in this chapter to facilitate the research and development in this important area.

1.1 Introduction

The world's increasing energy requirement, continuous and rapid depletion of fossil fuels, along with the alarming increase in the concentration of green house/polluting gases have driven large-scale research into the development of alternative and "greener" energy sources [1, 2]. Fuel cells, one of the energy-converting devices with a low/zero pollution emission, have generated a lot of interest among the scientific and engineering communities due to their higher energy densities and energy efficiencies compared to other current/conventional systems. DAFCs, using liquid and renewable alcohol fuel (e.g., methanol, ethanol), which is easily stored and transported and simplifies the fuel cell system, have been considered to be a favorable option in terms of both fuel usage and feed strategies [3–5]. However, two of the major challenges for DAFCs' commercialization are the insufficient catalytic activity/stability and high cost of the anode electrocatalysts. Currently, many studies have focused on this area, exploring new anode catalysts that can effectively enhance the alcohol electro-oxidation kinetics and reduce the cost. In this chapter, the development of DAFC anode catalysts in recent years is reviewed, mainly focusing on the five most active areas, namely, Pt-based catalysts, non-Pt catalysts, support materials for catalysts, surface structure controlling of catalysts, and theoretical modeling for fundamental understanding of catalytic activity and new catalyst design.

1.2 Pt-Based Catalysts

1.2.1 Alloy Catalysts

Until now, Pt has been the most effective metal catalyst for alcohol oxidation in acid medium. Unfortunately, for a pure platinum catalyst, the adsorption of CO, one of the intermediates in alcohol electro-oxidation, can occupy the reaction active sites, resulting in slow reaction kinetics [3, 6, 7]. To overcome this problem, extensive efforts have been devoted to promote the catalyst performance and decrease the catalyst cost by making Pt-based alloys with a second metal or even a third metal as the alloying component. The enhanced activity of these alloy materials when compared with pure Pt for alcohol oxidation has been attributed to a bifunctional mechanism and a ligand effect (electronic effect) [4, 6, 8–10]. It is widely recognized that the bifunctional mechanism involves the adsorption of oxygen-containing species on the foreign metal atoms at lower potentials, thereby promoting the oxidative removal of CO. An electronic model is also used to explain the enhancement effect. The alloying metal can modify Pt electronic structure, making Pt atoms less poisoned by CO or alcohol dissociative adsorption. Experiment results show that the catalytic activity of the Pt-based alloy catalysts is strongly dependent on the composition, structure, morphology, particle size, and alloyed degree.

In literature, the metal, such as Ru [10–13], Co [14], Sn [15–19], Bi [20], Mo [21, 22], Pd [23, 24], Cu [25], Ni [26, 27], etc., is usually introduced as alloying metal in Pt alloys. The addition of a second metal not only reduces the use of platinum, but can also enhance the activity of the catalysts for alcohol oxidation. Among these binary Pt-based alloy catalysts, Pt–Ru alloys are considered the most adapted for methanol oxidation, whereas in the case of ethanol electro-oxidation, Pt–Sn alloys seem to be the best ones. Generally, other binary Pt-based catalysts present a higher activity than that of Pt alone but lower than that of Pt–Ru and Pt–Sn. For example, Yang et al. [23] synthesized the PtPd alloy nanoparticles on Nafion–graphene film by using a simple electrochemical method. TEM images in Fig. 1.1 show that each PtPd particle is actually composed of many small nanoparticles. In electrochemical measurements, the

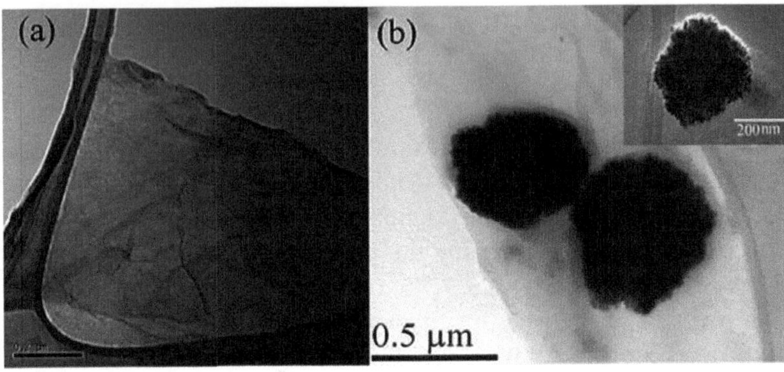

Figure 1.1 TEM images of (a) Nafion–graphene film and (b) bimetallic PtPd particles on the Nafion–graphene film. Reproduced with permission from Yang et al. [23]. Copyright © 2012 Royal Society of Chemistry.

PtPd alloy nanoparticles exhibited efficient electrocatalytic activity and stability toward ethanol oxidation in alkaline media. Moreover, this anode catalyst had a low onset potential at −0.73 V (vs. Ag/AgCl) and a high ratio of forward peak current density to backward peak current density (I_f/I_b) of 5.45, indicating that the catalyst had a good tolerance against poisoning by intermediates generated during ethanol electro-oxidation and was also effective in cleaving the C–C bond to achieve oxidation of ethanol to CO_2. Wang et al. [25] prepared uniform and well-dispersed $PtCu_2$ nanocrystals via one-pot co-reduction of H_2PtCl_6 and $Cu(acac)_2$ by oleylamine and polyvinyl pyrrolidone (PVP) in an autoclave at 180°C (Fig. 1.2a,b). The magnified TEM images indicate that the $PtCu_2$ nanocrystals exhibit concave surface with eight dendrites (Fig. 1.2c, d). On the basis of cyclic voltammetry and chronoamperometry, the concave dendritic $PtCu_2/C$ nanocrystals showed an exceptionally high activity and a strong poisoning resistance in methanol oxidation reaction. The enhanced catalytic activity could be attributed to the unique concave dendritic morphology of the bimetallic nanocrystals.

As described above, the Pt–Ru and Pt–Sn alloys are the best binary catalysts for the alcohol oxidation reaction. However, the Ru metal can be dissolved from the Pt–Ru catalyst surface under electrochemical conditions, leading to the performance degradation

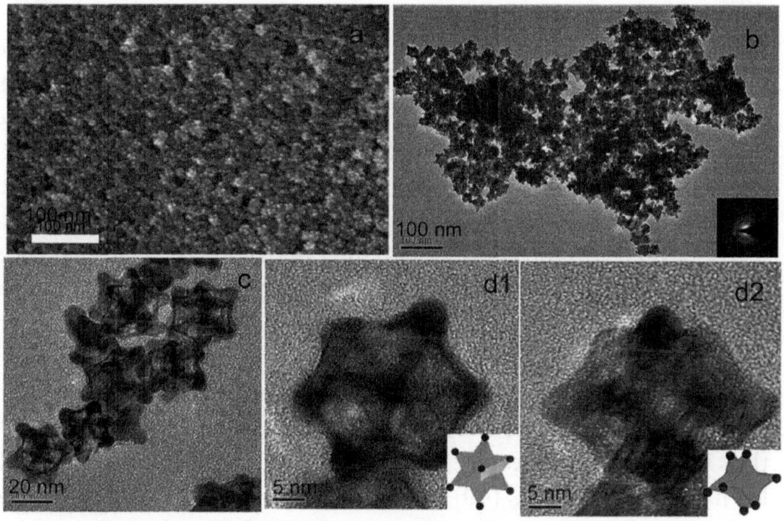

Figure 1.2 (a) FESEM image, (b) TEM image inset: SAED pattern of $PtCu_2$ nanocrystals, (c) magnified TEM image of concaved dendritic $PtCu_2$ nanocrystals, (d1, d2) HRTEM images of $PtCu_2$, the insets being the geometric models. Reproduced with permission from Wang et al. [25]. Copyright © 2014 Elsevier.

of fuel cell [8]. On the other hand, the complete electro-oxidation of ethanol from ethanol to CO_2 even using Pt–Sn catalyst does not easily occur at or near ambient temperatures. This is due to the difficulty in breaking C–C bond of ethanol, and hence, the major products in ethanol oxidation are usually CH_3CHO and CH_3COOH with a small amount of CO_2 (a few % or less) [28, 29]. Therefore, it is still an ongoing task to further improve the performance of the Pt–base anode catalysts. Adding the third metal into the binary Pt–alloy catalyst to form a ternary one seemed to be beneficial. The main advantage of the addition of a third metal to a bimetallic electrocatalyst is the reduction of the oxidation potential of small organic molecules, helping C–C breakage at low temperature, and then resulting in a complete oxidation of ethanol. So far, all the ternary Pt-based catalysts reported in literature, such as Pt–Ir–Sn [30], Pt–Sn–Mo [31], Pt–Ru–Sn [32], Pt–Re–Sn [33], Pt–Sn–Ni [34–36], Pt–Sn–Ce [37], Pt–Ru–Rh [38], Pt–Ru–Mo [39], etc., could always exhibit a better performance than Pt–Ru (Sn) catalysts for the ethanol oxidation reaction. It is noted that the ternary Pt–Sn–Ru catalyst with nominal Ru/Sn atomic ratio <1 seems to be the most promising anode material for the use in direct ethanol fuel cells. A suitable non-alloyed Ru to non-alloyed Sn ratio seems to be 0.4. On the Pt–Sn catalyst particle surface, there exists RuO_2–SnO_2 in aqueous solution. The interaction between RuO_2 and SnO_2 could promote the formation of hydroxyl species by dissociating water at a lower potential when compared to the Pt–Ru catalyst. Moreover, this interaction could also weaken the bonding between the hydroxyl species and the catalyst surface as compared with the bonding on Pt–Ru nanoparticles. The more weakly adsorbed hydroxyl species could further promote the electro-oxidation of adsorbed CO and acetaldehyde species on the active metal sites at a lower potential, thus improving the electrocatalytic performance. On the other hand, for higher Ru contents [40], as in the case of ternary catalyst of Pt–Sn–Ru (1:1:1) (the RuO_2/SnO_2 ratio in Pt–Sn–Ru (1:1:1) is twice than that in Pt–Sn–Ru (1:1:0.3)), RuO_2 could substitute SnO_2 in the interaction with Pt. The presence of RuO_2 on the particle surface of the Pt–Sn–Ru catalyst could decrease the active surface area of Pt particles. As a consequence, part of the noble metal becomes inactive due to the blocking of the Pt surface by the Ru oxide, leading to the inferior performance of the Pt–Sn–Ru/C catalyst with the atomic ratio (1:1:1) compared to Pt–Sn/C (1:1). Another important parameter affecting the ethanol oxidation activity of the Pt–Sn–Ru/C catalyst is the ratio of the Pt content and the oxide content ($SnO_2 + RuO_2$). It has to be pointed out that although the addition of Ru to Pt–Sn can enhance the ethanol oxidation activity of the catalyst, it does not modify the product distribution, suggesting that the ternary Pt–Sn–Ru catalyst is unable to activate the C–C bond cleavage. Therefore, basically the C–C bond scission remains the main problem in the ethanol oxidation reaction.

Ammam et al. [41–43] reported the synthesis, characterization, and ethanol oxidation activity of a series of binary Pt–Mn/C, ternary Pt–Mn–X/C (X = Fe, Co, Ni, Cu, Mo and Sn), and quaternary Pt–Mn–Cu–X/C (X = Fe, Co, Ni and Sn) and Pt–Mn–Mo–X/C (X = Fe, Co, Ni, Cu and Sn) alloy catalysts. In the case of quaternary Pt–Mn–Cu–X/C (X = Fe, Co, Ni, or Sn) and Pt–Mn–Mo–X/C (X = Fe, Co, Ni, Cu and Sn) catalysts, XRD results illustrated that the Pt–Mn–Cu–X/C (X = Fe, Co and Ni) and Pt–Mn–Mo–X/C (X = Fe, Co, Ni and Cu) alloys had been formed without significant free Mn, Cu, Mo, or X co–catalysts. For Pt–Mn–Cu–Sn/C and Pt–Mn–Mo–Sn/C, in addition to alloy formation, significant free Sn oxides were present in each catalyst. Cyclic voltammetry and chronoamperometry revealed that all quaternary alloys showed superior electrocatalytic activities toward ethanol oxidation when compared to the ternary precursor. A shift in the onset potential of ethanol electro-oxidation toward less positive values was also recorded with the quaternary alloys, demonstrating a facilitated oxidation with the quaternary alloys when compared to ternary alloy precursor. Overall, the incorporation of a fourth metal in the ternary Pt–Mn–Cu/C and Pt–Mn–Mo/C can lead to quaternary alloys with a better catalytic activity toward ethanol oxidation.

1.2.2 Oxide-Doped Pt Catalysts

During the synthesis process of Pt-based catalysts, many metal salts can be converted to metal oxides to form oxide-doped composites. These hybrid electrode materials, in which metal Pt is embedded in oxide, have been widely studied and demonstrated much improved electrocatalytic performance toward alcohol oxidation. These metal oxides include MnO_2 [44, 45], RuO_2 [46], CeO_2 [47–49], MoO_2 [50], IrO_2 [51], SnO_2 [52], TiO_2 [53, 54], WO_3 [55, 56], etc. For example, Huang et al. [45] reported a simple and cost-effective solvothermal treatment strategy for the synthesis of a ternary Pt/MnO_2/graphene sheets (Pt/MnO_2/GS) composite as an advanced electrocatalyst for high-performance DMFCs. It was claimed that in such a nanohybrid each component could provide unique and critical function to achieve optimized utilization of metallic platinum, allowing it to exhibit high electrocatalytic activity, good anti-poisoning ability and exceptional long-term stability, as well as outperforming the Pt/GS, Pt/XC–72 and Pt/MnO_2/XC–72 catalysts. Chu et al. [47] prepared a Pt/C–CeO_2 catalyst using a novel structural design of mesoporous CeO_2 (MP–CeO_2) coated with a certain amount of carbon from the β-cyclodextrin carbonization and then depositing the Pt nanoparticles through a fast and facile microwave-assisted polyol process. The results showed that the Pt/C–CeO_2 catalyst exhibited a significantly higher catalytic activity and stability for methanol electro-oxidation.

Yang et al. [55] synthesized carbon supported Pt–WO_3 nanoparticles by a microwave-assisted microemulsion method. In the synthesis, they found that the platinum precursor was uniformly impregnated into carbon pores when microwave irradiation was triggered, and the Pt nanoparticles preferentially nucleated at the WO_3 sites. Their method could produce a homogeneous distribution of nanoparticles on the carbon support with a uniform Pt:W ratio. Cyclic voltammetry and chronopotentiometry experiments demonstrated that a 1:1 ratio of Pt:W had the highest electrocatalytic activity for methanol oxidation. The enhanced electrocatalytic activity was attributed to both physical and chemical influence of the WO_3 component. More recently, Yan et al. [56] synthesized some WO_3 nanoparticles with an average diameter of 4.9 nm uniformly dispersed on porous carbonized resin (C–WO_3) through ionic exchange route, and these WO_3 nanoparticles were used as the catalyst support for Pt nanoparticles. Owing to the promotion effect of WO_3 on noble metal electrocatalyst, the Pt/C–WO_3 showed a higher electrocatalytic activity than the commercial Pt/C (TKK) for methanol oxidation at the same Pt loadings.

1.3 Non-Pt Catalysts

1.3.1 Pd-Based Catalysts

Because of the high price and global scarcity of platinum (Pt), partial or complete replacement of Pt metal has been an effective approach to the cost reduction [57]. In this regard, many efforts have been made to develop materials that may replace Pt as the anode catalyst in DAFCs. It has been expected that these materials would be better than normal alloy catalysts in terms of resistance to both CO and acid. Pd has been considered to be a promising alternative to Pt because Pd catalyst presents a higher electrocatalytic activity and better stability for alcohol electro-oxidation in alkaline media [58–61]. In addition, Pd is much more abundant on the earth and thus the catalyst cost can be significantly reduced [62, 63]. In recent years, various Pd catalysts, including nanoporous Pd [64, 65], Pd-based alloys [66–71], Pd nanoplates array [72, 73], Pd nanorods [74], Pd nanoflowers [75, 76], hollow Pd nanospheres [77], as well as core–shell nanostructures [78, 79], have been successfully explored that exhibit higher activity and stability for the alcohol oxidation reaction. For example, Sawangphruk et al. [64] fabricated the 3D ultraporous Pd nanocatalysts supported onto the surface of an rGO/CFP electrode by the cyclic voltammetry method. The electrodeposition reaction of Pd during the backward scan is $Pd^{2+} + 2e^- \rightarrow Pd$. On the other hand, during the forward scan, the stripping (oxidation) peak of Pd ($Pd \rightarrow Pd^{2+} + 2e^-$) was observed at −0.15 V vs. Ag/AgCl. It is believed that this stripping process should be a key leading to the ultraporous Pd located on the entire surface of rGO/CFP (Fig. 1.3a,b). The diameter of each Pd nanorod (Fig. 1.3b) has an average size of about 20 nm. The high-resolution TEM image (Fig. 1.3c) and electron diffraction pattern

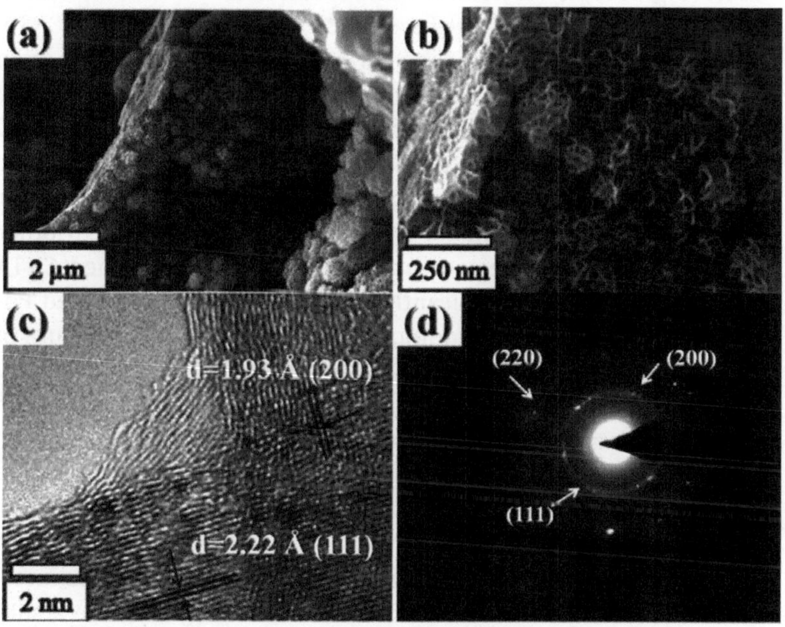

Figure 1.3 (a) Lower- and (b) higher-magnification SEM images of as-electrodeposited Pd on the rGO/CFP electrode, (c) HRTEM image, and (d) electron diffraction pattern of the as-electrodeposited Pd. Reproduced with permission from Sawangphruk et al. [64]. Copyright © 2013 Royal Society of Chemistry.

(Fig. 1.3d) indicate that the Pd catalyst can exhibit a typical face-centered cubic (fcc) lattice structure. The as-fabricated electrode of PdNCs/rGO/CFP could exhibit a higher catalytic activity toward the electro-oxidation of ethanol in alkaline media, higher poisoning tolerance to carbonaceous species, and also higher stability than other electrodes.

Hu et al. [76] reported Pd on carbon nanotube arrays (CNTA) with ends cut open by femtosecond laser as the catalyst of alcohol oxidation in alkaline media, as shown in Fig. 1.4. It can be seen that there are clearly some small ditch-like features on the surface of the end-opened CNTA, which were formed by the residual heat after femtosecond laser scanning (Fig. 1.4a,b). The depth of the ditches is about tens of micrometers, which is believed to be beneficial to the hydrophilic treatment of CNTA. However, the surface of uncut CNTA (Fig. 1.4c) is very smooth and featureless. The flower-like Pd particles are dispersed on the surface of CNTs within the CNTA. The Pd/$CNTA_{cut}$ electrode shows a better Pd particle distribution but similar Pd particle size as compared with Pd/CNTA electrode (Fig. 1.4d–f), which means that this end-opened electrode could have higher catalytic activity for ethanol oxidation reaction. Due to ditch-like features on the surface of the cut, Pd–CNTA could offer more active sites for ethanol oxidation when compared to other Pd structures. It was tested that the Pd-based end-opened CNTA electrode performed significantly better than the unprocessed Pd-based CNTA electrode. In addition, the aligned parallel structure of CNTA could also be beneficial to the improvement of the activity of the Pd/CNTA catalyst because of direct electron transfer, smaller ion diffusion resistance, and larger surface area of catalyst.

1.3.2 Other Non-noble Metal Catalysts

It is believed that developing a new generation of non-noble metal anode catalysts is essential for fuel cell technology because a substantial portion of the overall costs of current fuel cells is associated with the Pt catalysts. Many attempts have been directed toward the examination of the catalytic activity of cheap metals such as Ir-based [80, 81], Ni-based [82–85], and tungsten carbide [86] materials. Cao et al. [80] synthesized carbon-supported Ir_3Sn/C and Ir/C catalysts

Figure 1.4 SEM images of cut CNTA at low magnification (a), high magnification (b), uncut CNTA at high magnification (c), and TEM images of Pd/CNTA catalyst at high magnification (d), low magnification (e), Pd/CNTA cut catalyst at low magnification (f). Reprinted with permission from Hu and Chen [76]. Copyright © 2011 Elsevier.

with $NaBH_4$ as a reducing agent under the protection of ethylene glycol at room temperature. The electrochemical results indicated that the Ir-based catalysts showed superior electrocatalytic activity toward ethanol oxidation to Pt/C and Pt_3Sn/C catalysts, mainly at a low-potential region. The overall performance of Ir_3Sn/C was comparable to that of the Pt_3Sn/C catalyst, making it a promising alternative choice of anode catalyst for direct ethanol fuel cells. Ni-based catalysts as platinum-free alternatives have gained much attention in recent years and demonstrated high catalytic activity for the oxidation of methanol in alkaline media. The main advantages of Ni-based catalysts are related to their electrochemical stability and resistance to poisoning. And nickel is a versatile catalytic material due to its surface oxidation properties.

In order to improve the catalytic durability, many researches focus on the modification of Ni-based catalysts by incorporating complexes or nonmetallic elements into Ni either individually or collectively to form nickel complexes or metalloid amorphous alloy catalysts. Zhang et al. [82] designed a one-pot strategy to prepare an Ni–P alloy nanowire network embedded on reduced graphene oxide (RGO) via a facile solvothermal method. The hybrid electrode of the Ni–P nanowire/RGO composite exhibited enhanced electrocatalytic activity toward the oxidation of methanol in alkaline solution. Abdel Hameed et al. [83] also studied the electrocatalytic oxidation of methanol on Ni–P and Ni–Cu–P supported over commercial carbon electrodes in 0.1 M KOH solution. The presence of little amounts of nickel and copper together on the surface of commercial carbon increases the methanol oxidation current density compared to that containing nickel alone. The highest catalytic activity is achieved at catalysts formed at 90°C from electroless deposition solutions adjusted at pH 9.0. A mediated electron transfer mechanism is proposed on the basis of $Ni(OH)_2/NiOOH$ transformation. A better efficiency is noticed at Ni–Cu–P/C catalyst because of the lower α/γ-NiOOH redox contribution with a good stabilization of β/β nickel oxyhydroxide form. Tungsten carbide (WC) is recognized as a promising catalyst, having the resemble activity of platinum among interstitial carbides. Oh et al. [86] synthesized WC nanoparticles catalyst supported on platelet type–carbon nanofibers (p-CNFs) for alkaline direct ethanol fuel cells. The significant increase in the catalytic activity of WC/p-CNFs for ethanol oxidation is due to the synergistic effect between WC nanoparticles and the p-CNF supports and the structure effect of the p-CNFs.

1.4 Support Materials for Catalysts

Conventionally, highly conductive carbon blacks (CBs) with turbostratic structures and high surface areas, such as Vulcan XC–72R (Cabot Corp., 250 m^2 g^{-1}), Shawinigan (Chevron, 80 m^2 g^{-1}), Black Pearl 2000 (BP2000, Cabot Corp., 1500 m^2 g^{-1}), Ketjen Black (KB EC600JD & KB EC600J, Ketjen International, 1270 m^2 g^{-1} and 800 m^2 g^{-1}, respectively), and Denka Black (DB, Denka, 65 m^2 g^{-1}), are used as fuel cell electrocatalyst supports to ensure large electrochemical reaction surfaces. However, a multitude of materials other than CBs have also been investigated as catalyst supports for PEM fuel cells including DMFCs. In recent years, the focus has been shifted toward nanostructured supports as they enable faster electron transfer and high electrocatalytic activity. In general, these may be classified into two main categories: (i) carbon-based/carbonaceous supports and (ii) non-carbonaceous supports. The first category consists of carbon nanostructures like carbon nanotubes (CNTs), carbon nanofibers (CNFs), mesoporous carbons (MCs), and graphene. These nanostructured allotropes of carbon materials could meet the essential fuel cell support requirements such as high specific surface area, high electrical conductivity, and relatively good stability in acid and alkaline media. Normally, catalyst supports that consist of carbon nanomaterials with higher graphitic nature (e.g., CNTs and CNFs) are reported to be more stable. Regarding anode catalysis, the anode catalysts are normally supported on some high-surface-area materials, such as carbon particles, in order to achieve high dispersion and maximum utilization, as well as to avoid catalyst agglomeration during the fuel cell operation. It is believed that the rapid development of nanotechnology, especially in the area of the synthesis of carbon nanomaterials, will create more stable and active supported catalysts.

Generally, an ideal support should offer the following properties: (i) good electrical conductivity, (ii) good catalyst–support interaction, (iii) large surface area, (iv) mesoporous structure enabling the ionomer and polymer electrolyte to bring the catalyst nanoparticles close to the reactants, i.e., to maximize the triple-phase boundary

(TPB), (v) good water handling capability to avoid flooding, (vi) good corrosion resistance, and (vii) easy recovery of the catalyst. A strong interaction between the catalyst and the support not only improves catalyst efficiency and decreases catalyst loss but also governs the charge transfer process. Furthermore, the support can also enhance the catalyst performance and durability by reducing catalyst poisoning (e.g., CO, S). Therefore, the selection of support material is vital in determining the behavior, activity, longevity, and cost-effectiveness of the catalyst and then the overall fuel cell performance.

In order to achieve fine catalyst dispersion, high utilization, and stable nanoscale metallic particles, catalyst-supporting approaches are necessary. This is because compared to bulk metal catalysts, supported catalysts can show both higher activity and stability. As discussed above, carbon particles are frequently used as catalyst supports because of their relative stability in both acidic and basic media, good electric conductivity, and high specific surface area. Regarding DAFCs, in the past, several carbon materials have been tested as catalyst supports. In these anode catalysts, carbon support materials have shown a strong influence on the properties of supported noble metal catalysts, such as metal particle size, morphology, size distribution, alloyed degree, stability, and dispersion, affecting their performance in fuel cells through changing mass transport and catalyst layer electronic conductivity, electrochemical active area, and metal nanoparticle stability during the operation. Therefore, the optimization of carbon supports is very important in DAFC development.

In developing DAFC anode catalysts, the properties of a suitable carbon support, such as specific surface area, porosity, morphology, surface functional groups, electronic conductivity, corrosion resistance, etc., must be down-selected in order to achieve high-performing catalysts. In general, the properties of the carbon support materials have strong effects on the preparation procedures and performance of synthesized supported catalysts. Although considerable efforts have been made in the last decades to optimize the supporting strategies through both theoretical and experimental approaches, the physical and chemical origins of those effects are still not fully understood yet. In recent years, numerous studies have concentrated on understanding the effects of carbon supports and exploring new carbon supports.

In summary, electrocatalyst supports play a vital role in ascertaining the performance, durability, and cost of PEM fuel cells, including DAFCs [87, 88]. A myriad of nanostructured materials, including carbon nanostructures, metal oxides, conducting polymers, and many hybrid conjugates, have been exhaustively researched over the past few decades to improve the existing catalyst support materials. The following section will review the recent developments and investigations reported on various catalyst supports.

1.4.1 Nanostructured Carbon

Carbon blacks (CBs) (especially Vulcan XC–72) are the most commonly used supports for Pt-based catalysts of fuel cells in many studies and commercial applications [1, 3, 87]. They are usually prepared by the pyrolysis of hydrocarbons such as natural gas or oil fractions from petroleum processing. Normally, CB consists of near-spherical particles of graphite with an average diameter of less than 50 nm. These particles can coalesce into particle aggregates with polycrystallite structures of non-discrete 3D groupings which have an average diameter of around 250 nm. Each crystallite consists of several "turbostratic" layers with an interplanar spacing of 0.35–0.38 nm. The type of source material and the process of thermal decomposition in synthesizing CB largely determine the CB's morphology and particle size distribution. The high surface area (~250 $m^2\ g^{-1}$ for Vulcan XC–72), low cost, and easy availability of CBs can help reduce the overall cost of the fuel cell. Although widely used as catalyst support, CBs still suffer from problems such as (i) the presence of organo-sulphur impurities and (ii) deep micropores or recesses which can trap the catalyst nanoparticles, making them inaccessible to reactants, thus leading to reduced catalytic activity. The pore size and pore distribution also affect the interaction between Nafion® ionomer and the catalyst nanoparticles. Since the size of Nafion® micelles (>40 nm) is larger than the recesses in the CB, any metal nanoparticles in pores with diameter lower than the micelle size are not accessible to Nafion® and provide no contribution to the electrochemical activity. Furthermore, relatively to say, CB is thermochemically unstable. Thermochemical stability is required under acidic conditions of a typical DAFC, otherwise the corrosion of the carbon support could result in the disintegration of the catalyst layer.

The use of carbon materials as catalyst supports for precious metals has been rapidly increased in the last years because of the continuous advancements in fuel cells. The main requirements of suitable supports for fuel cell catalysts are high surface area, good electrical conductivity, suitable porosity to allow good reactant flux, and high stability in fuel cell environment. In general, CBs are used as catalyst support in low-temperature fuel cells. In fact, the high availability and low cost make CBs the most used support for fuel cell catalysts. The disadvantage of these CBs is the presence of a high amount of micropore, which may hinder the reactant flow. Moreover, these materials present low stability at temperatures higher than 100°C.

Recent studies have revealed that the physical properties of the carbon support can greatly affect the electrochemical properties of the fuel cell catalyst. It has been reported that carbon materials with both high surface area and good crystallinity can not only provide a high dispersion of Pt nanoparticles, but also facilitate electron transfer, resulting in better fuel cell performance [89, 90]. To further improve the catalyst performance, other types of advanced carbon materials have also been explored as fuel cell catalyst supports. The novel nonconventional carbon materials have attracted much interest as electrocatalyst supports because of their good electrical and mechanical properties and their versatility in pore size and pore distribution tailoring. These materials can also present a different morphology than carbon blacks both at the nanoscopic level in terms of their pore texture (e.g., mesopore carbon) and at the macroscopic level in terms of their form (e.g., microsphere). For example, studies on CNTs, mesoporous carbon, graphene, and CNF-supported catalysts have shown improved catalyst efficiency and higher electrochemical surface areas (ESA). These properties are highly desirable for reducing catalyst loadings, thereby lowering the overall fuel cell cost [88, 91]. The characteristics of some new carbon materials, metal dispersion, and the electrochemical activity of catalysts supported on these novel materials, when compared with those of catalysts supported on carbon blacks, will be reviewed in the following sections.

1.4.1.1 Carbon nanotubes

In recent years, a series of new nanostructured carbon materials have been explored as fuel cell catalyst supports. For example, the family of carbon nanotubes (CNTs) is the most well-known nanostructured carbon, which has shown very promising results in catalyst support for fuel cell applications due to their unique electrical and structural properties. The tubular structure of CNTs makes them unique among different forms of carbon materials, and they can be exploited as a kind of alternative material for catalyst support in fuel cells because of their high surface area, excellent electronic conductivity, and high chemical stability. Many studies explored CNTs as supports for DAFC catalysts in recent years, and the results showed that CNTs were superior to carbon blacks as catalyst supports for DAFCs [3, 87, 89, 92, 93]. The higher catalytic activity was attributed to the larger surface area of CNT architecture and the lower overpotential for alcohol oxidation. Therefore, CNTs appear to have promising potential as catalyst supports for DAFCs. It is believed that the high crystallinity of CNTs makes them highly conductive. The high surface area and amount of mesopores result in high metal dispersion and good reactant flux in tubular structure. Moreover, CNTs have a positive effect on Pt structure, resulting in a higher catalytic activity and a higher stability than carbon blacks. A challenge for the commercialization of CNTs is their higher cost compared to that of carbon blacks.

Normally, conventional CNTs are made of seamless cylinders of hexagonal carbon networks with forms of single-wall (SWCNTs) or multiwall carbon nanotubes (MWCNTs) [87]. An SWCNT is a single graphene sheet rolled into a cylinder, while an MWCNT consists of several coaxially arranged graphene sheets rolled into a cylinder. The graphene sheets are stacked parallel to the growth axis of CNTs, and their spacing is typically 0.34 nm. Another form of CNT is stacked-cup carbon nanotubes (SCCNTs) consisting of truncated conical graphene layers. Multiwalled nanotubes may exhibit high degree of uniformity of internal diameter of single tubes, but with broad pore-size distribution in the micropore and mesopore ranges. When CNTs are used as the catalyst supports, their typical dimensions are as follows: outer diameter 10–50 nm, inside diameter 3–15 nm, and length from 10 to 50 μm. According to theoretical predictions, SWCNTs could be either metallic or semiconducting depending on the tube diameter and helicity [94]. For MWCNTs, scanning tunneling spectroscopy (STS) measurements indicated that the conduction was mainly due to the outer shell, which was usually much larger than SWCNTs [95]. Therefore, MWCNTs should have a relatively high electrical conductivity. An important aspect of the MWCNTs is their high surface area for subsequent metal deposition.

As for CNT synthesis, several methods can be used such as arc discharge, laser ablation, chemical vapor deposition (CVD), and plasma-enhanced chemical vapor deposition (PECVD) [3, 87]. For the synthesis of CNT-supported catalysts, several methods such as impregnation, ultrasound, polyol and microwave-assisted polyol, sputter deposition, precipitation, colloidal, ion-exchange, and electrochemical deposition (either pulsed or continuous) can be used for the deposition of electrocatalyst nanoparticles on CNTs [1, 96]. However, pristine CNTs are chemically inert with a hydrophobic surface, which makes it difficult for them to attach to metal nanoparticles. Thus, functionalization to introduce surface oxygen groups (using strong acids like HNO_3, H_2SO_4, etc.) and others (using ionic liquids, aromatic molecules, polymer, etc.) have been performed commonly in order to make the surface more hydrophilic and improve the catalyst support interaction. After CNTs were discovered by Iijima in 1991 [97], the first acid-functionalized CNTs for deposition of the catalyst nanoparticles were reported in the late 1990s [98]. Since then, the process has been continuously employed to obtain better dispersion of metal catalyst nanoparticles, better size control and distribution, and also selective morphologies for application in PEM fuel cells, including DAFCs. In order to improve the dispersion of catalyst nanoparticles (NPs) on CNTs, efforts have been focused on functionalizing CNTs, such as controlling their hydrophobic properties and introducing active sites on their surface [99–103]. For example, Li et al. [104] successfully synthesized a

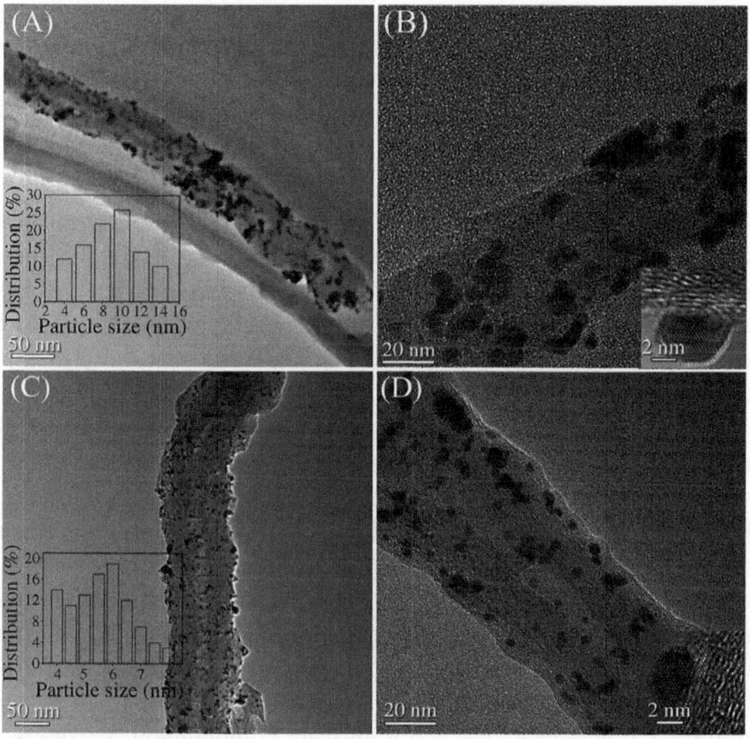

Figure 1.5 TEM images of Pd@CNTs (A and B) and Pd@IL (Cl^-)-CNTs (C and D). Inset: size distribution histograms and HRTEM images of the crystal structure of Pd NPs in detail on CNTs and IL (Cl^-)-CNTs. Reprinted with permission from Li et al. [104]. Copyright © 2013 John Wiley and Sons.

Pd@IL(Cl^-)-CNT catalyst (IL: ionic liquid) for direct ethanol alkaline fuel cells. The strategy of IL functionalization of CNTs could be useful for preparing other carbonaceous carrier supports to enhance the dispersivity, stability, and catalytic performance of metal NPs as well. The TEM images of Pd@CNTs (Fig. 1.5A,B) show that Pd NPs are randomly dispersed on CNTs and some of them appeared to aggregate with a wide size distribution (the size of NPs varies from 4 to 14 nm). The quasi-spherical-shaped Pd NPs are equably attached on the surface of IL(Cl^-)-CNTs (Fig. 1.5C,D) and possess more perfect crystal structure than that of the Pd@CNTs (HRTEM). Surface-modified CNTs were also used to support a wide variety of mono, binary (e.g., Pt–Ru, Pt–Co, Pt–Fe), as well as ternary (e.g., Pb–Pt–Cu, Pt–Ru–Ni, Pt–Ru–Ir) catalyst systems using both noble and non-precious metals [102, 105–111].

Normally, catalysts supported by CNTs with different morphologies and pretreated through different procedures can have different fuel cell performance. Tokarz et al. [112] synthesized some Pt–Ru/CNTs anode catalysts by supporting catalytic metal particles on CNTs with different morphologies and subjected to different pretreatment, and found that the catalyst with best performance both in DMFC and in H_2–PEMFC applications is that supported by thin and entangled CNTs graphitized at high temperature. This might be because this support could yield a catalyst giving the optimum combination of metal utilization and transport properties of the resulting catalyst layer. Furthermore, the improved conductivity of the graphitized CNTs could also play a role. The fuel cell with this best CNT-based catalyst matched the performance of cell prepared using the reference Vulcan-supported catalyst despite the much thicker anode catalyst layer of the former relative to the latter, demonstrating the usefulness of the CNTs for the preparation of fuel cell catalyst layers with improved transport properties. From the TEM images of the catalyst before test (Fig. 1.6a,c,e), it can be seen that the catalyst prepared is not entirely homogeneous on the tens-of-nanometers scale. TEM images of the NT templ (i.e., thick and disentangled CNTs) and the NT graph (i.e., thin and entangled CNTs graphitized at 2500°C) catalysts sampled from MEAs after fuel cell test (Fig. 1.6b,d) reveal a metal particle agglomeration phenomenon caused by fuel cell operation. An obvious feature on the images of the used catalysts is that the small Pt–Ru particles could be detached from the supports with the recast Nafion® (Fig. 1.6d).

Actually, CNT synthesis, metal loading, and electrode preparation based on CNT supports still face some challenges, especially when applied to fuel cells. As mentioned above, CNTs are usually synthesized by carbon-arc discharge, laser ablation of carbon, or CVD (typically on catalytic particles). These synthetic methods have their limitations in terms of large-scale production and cost-effectiveness. Their harsh synthetic conditions and low production yields are also major disadvantages. It is necessary to further develop industrial-level large-scale production of CNTs to meet the needs of all the possible applications, including the one in fuel cells.

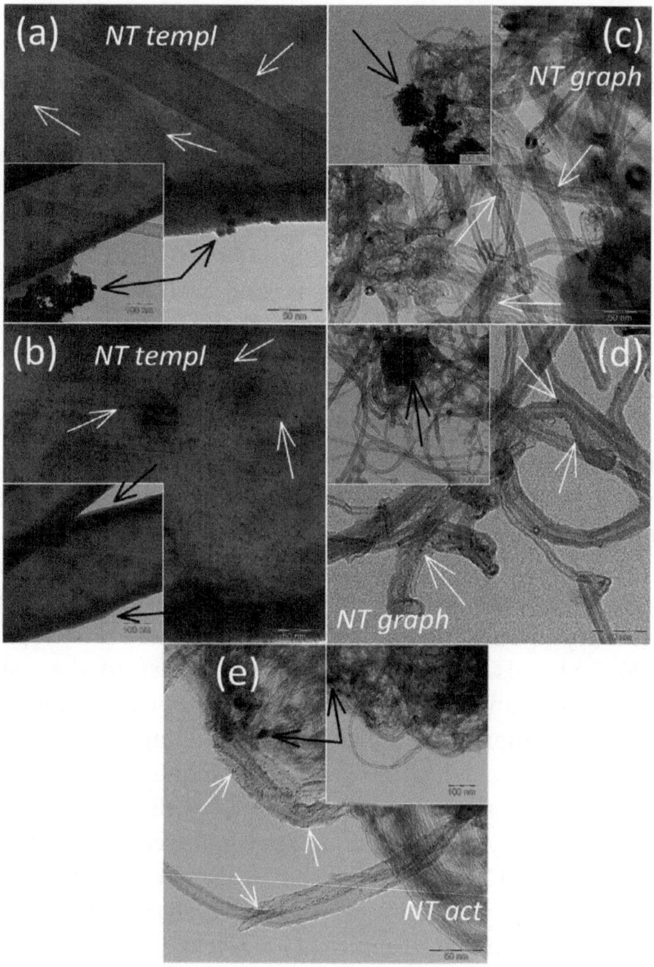

Figure 1.6 TEM images of pristine Pt–Ru/CNT catalysts (a, c, e) and Pt–Ru/CNT catalysts having worked in fuel cell (b, d). Type of CNT support indicated in images. Insets show lower magnification. Reprinted with permission from Tokarz et al. [112]. Copyright © 2013 Elsevier.

1.4.1.2 Carbon nanofiber

Carbon nanofibers (CNFs) can offer the advantages of an immobile catalyst and a short diffusion distance. Another advantage of fibrous catalysts is their low resistance to flow of liquid and gases through a bundle of fibers. Thus, they can be used as an attractive alternative support for catalysts in fuel cell. Therefore, activated carbon fibers represent a novel kind of porous material, with high surface area (>1000 m^2 g^{-1}), and the presence of a lot of functional groups on the surface. Depending on the orientation of the nanofibers with respect to the growth axis, CNFs can be classified into three types: (i) ribbon-like, (ii) platelet, and (iii) herringbone (or stacked-cup). Herringbone CNFs are known to have intermediate characteristics between parallel and platelet types, and their supported catalyst could exhibit higher catalytic activity than the parallel and better durability than the platelet forms. Unlike conventional graphite materials and nanotubes where the basal plane is exposed, in the structure of CNFs, only the edge regions are exposed [87, 113]. The main difference between nanotubes and nanofibers is the lack of a hollow cavity in the latter. Owing to their peculiar structure, CNFs are mainly used as catalytic supports without any pretreatment [114]. For example, platelets and herringbone structures present potentially reactive groups for metal anchoring. It is well known that the key to successful utilization of the catalyst nanoparticles in PEM fuel cells is their simultaneous access to the reactants (hydrogen, methanol, oxygen, etc.), the electron-conducting medium (e.g., carbon paper), and the proton-conducting medium (usually Nafion), in a so called 'triple region'. A unique approach that has been adopted by several groups for solving these problems is to grow CNFs directly on the carbon paper, followed by electrocatalyst deposition [115, 116]. In recent years, CNFs have generated intense interest in their application as DAFCs catalyst support material because of their unique structure [117–121].

Feng et al. [119] fabricated the CeO_2 nanoparticles which were embedded in the CNFs by electrospinning, and used as a support for PtRu nanoparticle catalyst for methanol oxidation. In their work, the parameters affecting the performance of the catalyst layer structure on a glassy carbon electrode (GCE), such as the catalyst loading and the Nafion® content in the catalyst layer, were also investigated. The catalyst layer prepared on the GCE with the nanofiber-supported catalyst showed a different activity from that of the nanoparticle catalyst, which was strongly dependent on both the loading of the catalyst and the Nafion® content. There was an

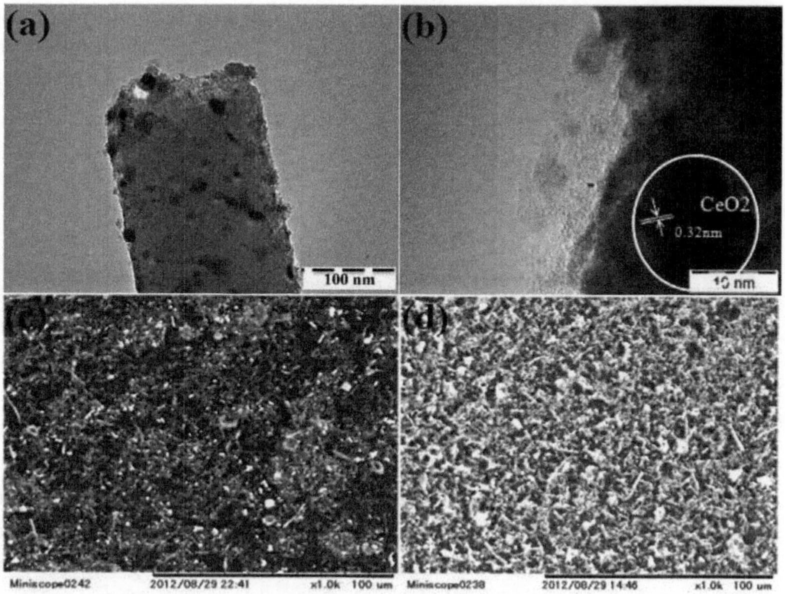

Figure 1.7 TEM images of the prepared PtRu/(C–CeO_2)NF at (a) low magnification and (b) high magnification. Surface morphology of the PtRu/(C–CeO_2)NF catalyst layer on the GCE using different catalyst loadings: (c) 0.25 and (d) 0.45 mg cm^{-2}. Reprinted with permission from Feng et al. [119]. Copyright © 2013 Elsevier.

optimum Nafion® content, 20 wt%, in the catalyst layer on the GCE for producing a homogenous layer structure. The homogenous layer structure with the nanofiber was obtained at a relatively high catalyst loading between 0.5 and 1.5 mg cm^{-2}. Under the optimum conditions, the catalyst layer with PtRu/(C–CeO_2)NF as catalyst showed three times higher mass activity for methanol oxidation and durability than that of the commercial PtRu/C catalyst due to the incorporation of CeO_2 into the catalyst layer. As shown in Fig. 1.7a, the nanofiber morphology is obvious in the image and the small particles dispersed on the support nanofiber are the PtRu nanoparticles. The big particles that are around 15 nm in diameter in the nanofibers were CeO_2 nanoparticles, as confirmed by measuring the interplanar spacing of 0.32 nm that is in agreement with the CeO_2 (111) facet distance (Fig. 1.7b). At the low catalyst loading (Fig. 1.7c), the heterogenous catalyst layer on the entire surface of the electrode can be formed; therefore, the electrode was not efficiently used. By increasing the catalyst loading, the entire surface of the electrode is homogeneously covered with the nanofiber catalyst (Fig. 1.7d).

1.4.1.3 Mesoporous carbon

According to the International Union of Pure and Applied Chemistry (IUPAC), pores are classified, depending on their width, as micropores (<2 nm), mesopores (2–50 nm), and macropores (>50 nm) [87]. Generally, carbon blacks have high specific surface areas but are mostly composed of micropores less than 1 nm and are therefore very difficult to access. The presence of micropores in a carbon material is not beneficial to its usage as catalyst support. When the average diameter of the pores is less than 2 nm, supply of a fuel to the catalyst layer may not be smooth and efficient, limiting the activity of the catalyst. Moreover, it is known that micropores of amorphous carbon particles are poorly connected. Compared to carbon blacks, mesoporous carbons (MCs) normally have higher surface area and lower amount or absence of micropores. In a mesoporous carbon-supported catalyst, the metal catalyst particles can be distributed and supported on the surface or in pores of the mesoporous carbon. A large mesopore surface area, particularly with pore size >20 nm, can give rise to a high dispersion of Pt particles, leading to a large effective surface area of Pt with a high catalytic activity. In addition, the mesoporous structure can facilitate a smooth mass transportation.

In general, MCs can be classified into two categories on the basis of their final structure and method of preparation: (i) ordered mesoporous carbons (OMC) and (ii) disordered mesoporous carbon (DOMC) [100, 122]. OMCs can be synthesized using ordered mesoporous silica templates [3, 87, 123]. The synthesis involves infiltration of the pores of the template with an appropriate carbon precursor (furfuryl alcohol, sucrose, acenaphthene and mesophase pitch, etc.), its carbonization, and subsequent template removal. The properties of the resultant carbon are strongly dependent on the structure of the template. Normally, MCM-48, SBA-1, and SBA-15 silicas can be successfully used to synthesize carbons with cubic or hexagonal frameworks, narrow mesopore-size distributions, high nitrogen BET specific surface areas (up to 1800 $m^2\ g^{-1}$), and large pore volumes. In literature, a wide range of applications have been explored for OMC, including Li–ion batteries and fuel cells [122]. They have also been widely studied as catalyst support materials for DAFCs [123–126]. It is commonly accepted that the structure of the carbon support is an important factor affecting the performance of the electrocatalyst. It can affect the accessibility of the reactants to the catalytic site and also the removal of the products. The interesting morphological structure of MC, with large surface area and monodispersed mesospheres connected three-dimensionally, could facilitate the diffusion of reactants and by-products, making them very attractive catalyst support materials. MC is also known to have some surface oxygen groups which are considered to have beneficial effects in improving the interaction between the metal catalyst and the carbon support [127].

Generally, a high-performing DAFC anode requires an efficient nanoscale three-phase reaction zone, in which the electrochemical reactions occur on the surface of the metal nanoparticle involving electron and proton transport. In addition, some efficient transport passways for both liquid-phase reactants (e.g., CH_3OH, C_2H_5OH) and the gas-phase product (CO_2) are also needed. Too many small micropores in carbon supports can decrease catalyst utilization because the mass transport of reactants and products is poor in these micropores. However, when the macroporous size is larger than 50 nm, the surface area will become small and the electrical resistance will increase. Mesoporous carbons with tunable pore sizes in the range of 2–50 nm are attractive for use as catalyst supports and have the potential to enhance both the dispersion and utilization of metal catalysts. Therefore, MCs have recently received great attention because of their potential usage as catalytic supports in fuel cell electrodes. Their higher performance was considered to be not only due to the higher surface areas and larger pore volumes, which allowed a higher degree of catalyst dispersion, but also due to highly integrated interconnected pore systems with periodic order, which allowed efficient transport of reactants and products [87, 88, 128].

1.4.1.4 Graphene

High conductive graphene is an atomically thin sheet of hexagonally arranged carbon atoms, which has attracted great interest since its discovery. It is widely studied for various applications, including fuel cell catalyst support [129, 130]. The use of graphene and its oxide (graphene oxide–GO) in fuel cells seemed limited to that of catalyst support, but it was also explored (i) as a material for conducting composite membranes with polymers and (ii) as bipolar plate material [131].

Graphene has aroused interest as potential support owing its high electron transfer rate, large surface area, and high conductivity [132–138]. The 2D planar structure of the graphene sheet allows both the edge planes and the basal planes to interact with the catalyst nanoparticles. The rippled but planar sheet structure can also provide a very high surface area for attaching catalyst nanoparticles. Liu et al. [132] fabricated some nanocomposite films of platinum nanoparticles supported on an expandable graphene sheet (Pt/EGS) which were coated on conductive indium tin oxide glass electrodes via a "green" electrochemical synthetic route involving a series of electrochemical processes. The catalytic activity and stability of the Pt/EGS film for the oxidation of methanol were evaluated through cyclic voltammetry and chronoamperometry tests. The Pt nanoparticles in the Pt/EGS nanocomposite film were found to be uniformly distributed on the EGS. The as-synthesized Pt/EGS nanocomposite exhibited high catalytic activity and good stability for the oxidation of methanol, which might be attributed to its excellent electrical conductivity and the high specific surface area of the graphene sheet catalyst support. From Fig. 1.8a, it can be seen that the expandable graphene oxide (EGO) film is made up of many stacked EGO layers with numerous edges. In the electrophoretic deposition process, the thickness of the EGO film could be easily controlled by adjusting the concentration of the EGO colloid solution and deposition time. Figure 1.8b reveals that the surface of EGS has no substantial difference from that of the EGO film, although more edges and fractures are visible in the former case. Figure 1.8c indicates that the Pt/EGS composite can be successfully synthesized via a controlled electrochemical reduction

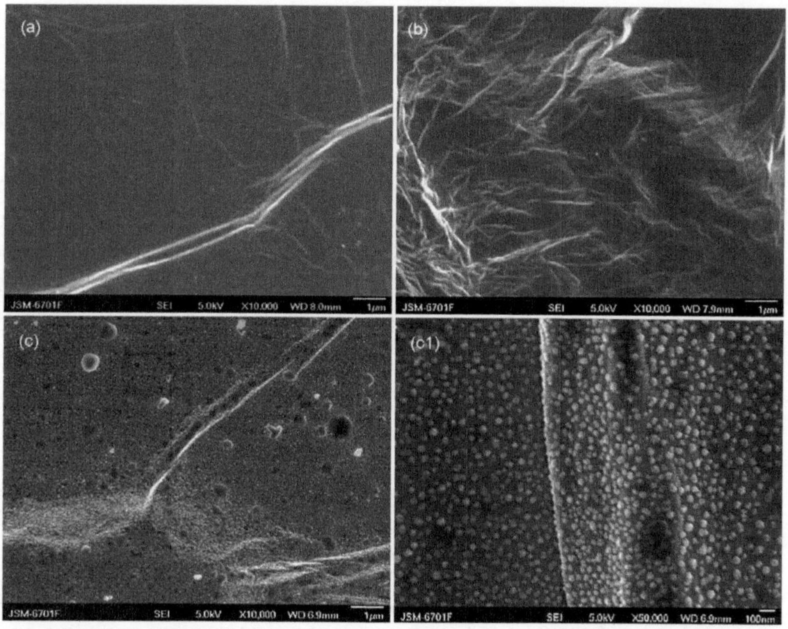

Figure 1.8 FE-SEM images of EGO (a), EGS (b), and Pt/EGS (c and c1) at increasing magnifications. Reprinted with permission from Liu et al. [132]. Copyright © 2010 Elsevier.

of $PtCl_6^{2-}$, where Pt nanoparticles are uniformly distributed on the EGS surface, and Pt nanoclusters on the EGS film are not severely aggregated (Fig. 1.8c1), and the Pt nanoparticles with a uniform diameter of about 15 nm are well dispersed on the EGS surface.

Yao et al. [135] reported a novel method for assembling reduced graphene oxide (RGO) and Pt nanoparticles on a carbon fiber (CF) electrode to form a stable Pt nanoparticle–RGO–Pt nanoparticle–RGO/CF multiple junction for electrocatalyst application. Pt nanoparticles are uniformly deposited on the surface of each RGO sheet, performing an alternative covering structure of RGO and Pt nanoparticle multilayer on the CF electrode to obtain a novel three-dimensional (3D), multilayered Pt/RGO-modified CF electrode (N-Pt/RGO/CF). Experimental results demonstrated that the prepared N-Pt/RGO/CF electrode showed better electrochemical properties and enhanced electrocatalytic activity toward methanol electro-oxidation in an alkaline medium when compared with the Pt/RGO/CF electrode without layer-by-layer structure or the Pt/CF electrode without RGO. This enhancement might be due to the unique 3D pore structure of N-Pt/RGO/CF and the good electron transport property of RGO in the composite electrode.

Zhong et al. [138] reported a facile and effective ultrasonication approach to noncovalently functionalize graphene with copper phthalocyanine tetrasulfonic acid tetrasodium salt (TSCuPc), and the functionalized graphene that was formed was used as a promising catalyst support for Pt nanoparticles. The results demonstrated that Pt nanoparticles are homogeneously deposited on the surface of TSCuPc functionalized graphene, and their dispersivity and electrochemical active surface area (ECSA) are obviously enhanced. Studies of cyclic voltammetry and chronoamperometry indicated that the as-prepared Pt/TSCuPc–graphene catalyst could exhibit both much higher electrocatalytic activity and stability than the Pt/graphene and commercial Pt/C catalysts for methanol oxidation.

In addition, graphene could also be used as both a reductant in synthesis of MnO_2 [45] and an ideal substrate for growing and anchoring Pt nanoparticles with a very narrow size distribution. Most importantly, the rationally designed Pt/MnO_2/GS ternary electrocatalyst exhibited an ultrahigh electrocatalytic activity, good anti-poisoning ability, and exceptional long-term stability, which significantly outperformed the Pt/GS, Pt/XC–72, and Pt/MnO_2/XC–72 catalysts. It was believed that this design concept could be extended to the fabrication of other novel nanocomposites containing noble metals, such as Pd, Au, and Ag, probably opening up a brand new avenue for a large spectrum of device applications.

Graphene oxide (GO) has also drawn some considerable interest and attention. Although GO has lower conductivity (two to three orders of magnitude lower than graphene), it offers a different set of properties (hydrophilicity, high mechanical strength, chemical "tunability") when compared to graphene, which makes it suitable for a wide range of different applications. Moreover, the variable oxygen content of GO enables tunable electronic conductivity for various applications. The use of GO as catalyst support material in PEM fuel cells, including DAFCs, is one of the latest applications which have shown promising results [139–144]. It is obvious that oxygen groups introduced into the graphene structure during the preparation of GO can create defect sites on surface as well as edge planes. These defect sites can act as nucleation centers and anchoring sites for growth of metal nanoparticles.

1.4.2 Conducting Polymers

Normally, most of the common conjugated organic polymers are either electronic insulators or semiconductors. However, recently new polymers have been discovered, which have considerable electronic conductivities. These conducting polymers (CPs) are conjugated polymers having an extended p-orbital system through which electrons can move from one end of the polymer chain to the other [145]. The maximum conductivity so far obtained for this kind of CPs is about a single order of magnitude below that of metallic copper. The conduction within the conducting polymers is provided by delocalized π-electrons along the polymer chain.

Many CPs have porous structure and high surface area, which can promote their use as supporting materials for the development of new catalytic and electrocatalytic materials. The relatively high electric conductivity of some polymers makes it possible to shuttle the electrons through polymer chains between the electrode and dispersed metal particles to promote the electrocatalytic reaction. Because of their high accessible surface area, low resistance, and high stability, CPs have received much attention and have been investigated as carbon-substitute supports for fuel cell catalysts [146, 147]. Given their conductive and stable three-dimensional structure, CPs can act as suitable supports for low-temperature fuel cell catalysts. As recognized, polymer/metal–nanoparticle composites allow a facile flow of electronic charges through the polymer matrix during the electrochemical process. Additionally, electrical conducting polymers provide a low ohmic drop for the electron transfer between the metal catalyst and the substrates. Therefore, CPs with porous structures and high surface areas have been employed as a matrix to incorporate noble metal catalysts for some electrochemical reactions of importance for application in fuel cells such as hydrogen and methanol oxidation and oxygen reduction reactions. The main reason for incorporating metallic particles into porous polymeric matrixes is to increase the specific area of the active surface and thereby improve catalytic efficiency. Another reason is the higher tolerance of polymer-supported platinum particles to poisoning induced by the adsorption of CO species. For conventional carbon-supported platinum catalysts, they are easily poisoned by the strong adsorption of CO present in H_2 or of CO-like intermediate species formed during alcohol oxidation, which block the active Pt sites, lowering its electrocatalytic activity [3, 7]. Furthermore, CPs are not only electron-conducting but also proton-conducting materials, so they can replace Nafion® in the catalyst layer of the fuel cell electrode and provide enhanced performance [148]. In this sense, only a two-phase boundary is necessary for electron and ion transfer during reactions in fuel cells rather than the three-phase boundary required when carbon is used as the support.

The most common CPs used as fuel cell catalyst supports are conjugated polymers with heteroatoms in the main chain (heterocyclic polymers), including polyaniline (PAni) [149, 150], polypyrrole (PPy) [151], polythiophene (PTh) [152], poly(o-phenylenediamine) (PoPD) [153], poly(diallyldimethylammonium chloride) (PDDA) [154], and poly(2-amino-5-mercapto-1,3,4-thiadiazole) (PAMT) [155]. PoPD is one of these kinds of polymers that is highly aromatic and contains 2,3-diaminophenazine repeating unit. Electro-oxidation of methanol on platinum particles dispersed into PoPD was studied extensively. Habibi et al. [153] synthesized a PoPD film on the carbon ceramic electrode (CCE) surface and then platinum nanoparticles were incorporated into the polymer matrix, and this electrode was used for ethanol electro-oxidation in H_2SO_4 solution. The results demonstrate that the Pt particles are uniformly distributed onto the polymer surface and a small amount of them are agglomerated in the Pt|PoPD|CCE compared to Pt|CCE. It is confirmed that the polymer acts as a protective matrix and prevents nanoparticles aggregation after their formation. The electrochemical tests show that the Pt|PoPD|CCE exhibits higher mass activity for ethanol oxidation reaction than Pt|CCE. The kinetic parameters of ethanol oxidation on the Pt|PoPD|CCE, i.e., reaction orders for ethanol and H^+, Tafel slope, and activation energy (E_a), were also obtained.

Moreover, some copolymer composites such as poly(3,4-ethylenedioxythiophene)-poly(styrene sulfonic acid) or PEDOT–PSS [156, 157], Polyaniline-polypyrrole (PAni–PPy) [158], and polyaniline-poly(styrene sulfonate) or PAni–PSS [159] were also explored as the catalyst supports, and the results were quite promising. For example, Arbizzani et al. [156] synthesized the PEDOT–PSS composites with different (3,4-ethylenedioxythiophene):styrene-4-sulfonate (EDOT:SS) molar ratios and prepared PEDOT–PSS/PtRu by both electrochemical and chemical deposition of PtRu. It was found that the specific catalytic activities of PtRu electrochemically deposited on the PEDOT–PSS supports are double the catalytic activity found for PtRu supported on Vulcan commercial carbon. Jia et al. [159] prepared the PAni–PSS hydrogel via supramolecular self-assembly and then fabricated the PAni–PSS hydrogel–Pt catalyst by the electrochemical deposition of Pt particles. The electrochemical measurements show that the PAni–PSS hydrogel–Pt catalyst has preferable electrocatalytic performance toward the oxidation of methanol than Pt particles supported on PAni–PSS colloid.

In summary, conducting polymers have the potential to be developed as high-performing catalyst supports in PEM fuel cells including DAFC because they are gas and water permeable and conduct both protons and electrons, as well as because of their other beneficial properties.

1.4.3 Hybrid Supports

Catalytic metal nanoparticles supported on hybrid support systems consisting of a combination of two different support materials have also been extensively explored and developed [160]. In such a hybrid support material, one serves as the primary material and the other as the secondary one. For example, the use of metal oxides and carbides with carbon nanostructures as primary as well as secondary supports has attracted attention [161–165]. SnO_2 nanowires grown onto carbon paper could be used as support for platinum or other precious metal nanoparticles to form hybrid-material supported catalysts [161]. Wang et al. [164] synthesized ordered a mesoporous tungsten carbide/carbon (WC/OMC) composite by combing the hydrothermal reaction and the hard template

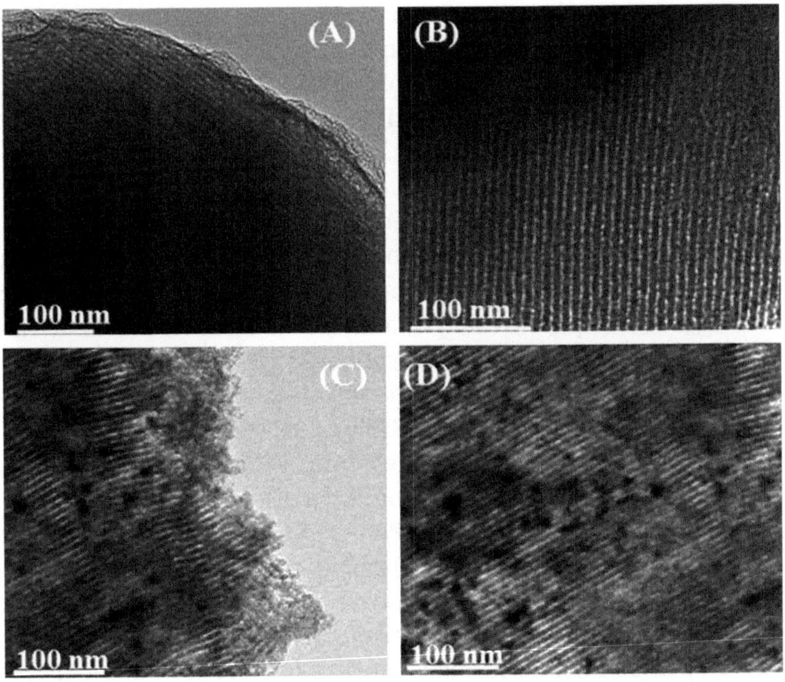

Figure 1.9 TEM images of OMC (A and B) and WC/OMC (C and D). Reprinted with permission from Wang et al. [164]. Copyright © 2014 Elsevier.

method. In this process, ammonium metagungstate (AMT) was used as tungsten precursor, glucose as the carbon source, and the ordered mesoporous silica (SBA-15) as the hard template. Thereafter, Pt nanoparticles were deposited on the as-prepared WC/OMC. The TEM image of OMC demonstrates the formation of a highly ordered 2D hexagonal mesostructure by the parallely arranged channels (Fig. 1.9A,B). When AMT was introduced as the tungsten precursor, it was obvious that the ordered structure was still maintained (Fig. 1.9C,D). The combination of high surface area, ordered mesopores, the synergistic effect, and CO-tolerant ability of WC and well-dispersed Pt nanoparticles provide Pt@WC/OMC with high activity, desirable stability, and CO tolerance toward methanol electro-oxidation. Compared with the commercial PtRu@C catalyst, Pt@WC/OMC demonstrates longer stability as well as higher specific mass activity for methanol electro-oxidation. Experimental results have proved that Pt@WC/OMC is a promising anode catalyst for DMFC applications. These significant improvements are of vital importance for the identification and development of low noble metal catalysts for DMFC applications. Finally, it is worth noting that the present work could also render a universal approach for the preparation of other ordered mesoporous composite materials for various important applications.

Karakoti et al. [165] used a mixed oxide system of ceria and titania as support for 2–5 nm Pt nanoparticles to study the effect of synthesis technique on the catalytic activity of Pt. The particle size and shape of Pt nanoparticles synthesized by thermal reduction (Fig. 1.10a–c) were identical to the calcined sample. The highly dispersed nature of Pt nanoparticles on the surface of oxide prevented further agglomeration during the calcinations. However, the possibility of support effect in pinning the coarsening of Pt nanoparticles cannot be ruled out. Surprisingly, no indication of a top oxide coating was evidenced by the TEM even though the samples showed a high degree of oxidized platinum.

On the other hand, conducting polymers deposited on carbon substrates are very favorable and attractive hybrid supports for catalyst particles, and some promising results were recently reported [152, 154, 166–168]. Wei et al. [167] fabricated a novel Pt-based electrocatalyst using an MWCNT-supported manganese oxide and poly(3,4-ethylenedioxythiophene) (PEDOT) nanocomposite (MnO_x–PEDOT–MWCNTs) as catalyst support for Pt nanoparticles. The morphology and size of the synthesized materials were analyzed by TEM (Fig. 1.11). The results demonstrate that the PEDOT film is homogeneously coated on the surface of MWCNTs, and the nanotubular morphology is retained in this composite (Fig. 1.11a,b). Interestingly, the inclusion of the MnO_x in the PEDOT–MWCNT composite matrix generated different microstructural features (Fig. 1.11c). The MnO_x component in the composite are MnO_x nanowire networks, which are closely combined with the PEDOT layer and homogeneously and densely attached to the surface of MWCNTs. The dispersion of Pt nanoparticles on MWCNTs is characterized by a poor distribution with a large number of aggregates (Fig. 1.11d). The average particle size of Pt nanoparticles

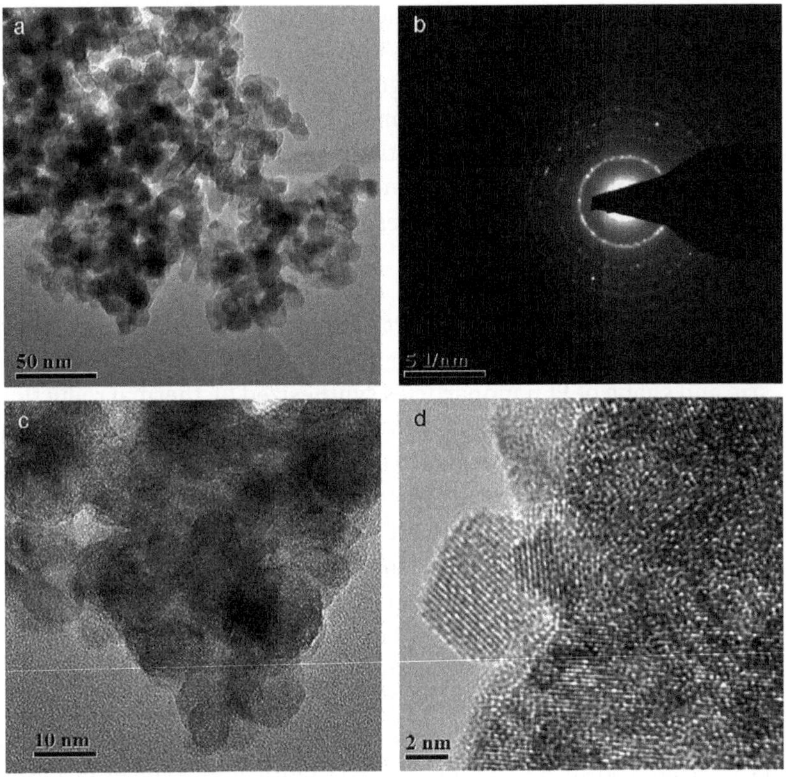

Figure 1.10 TEM images of sample PTCT over ceria–titania mixed oxide support: (a) low-magnification image; (b) selected area electron diffraction confirming the polycrystalline nature of the mixed oxide; (c) high magnification image depicting that the particle size of the base oxide is unaltered; (d) high magnification image depicting 2–5 nm platinum (111) nanoparticles. Reprinted with permission from Karakoti et al. [165]. Copyright © 2010 Elsevier.

is 4.75 ± 0.5 nm. It is likely due to the non-uniform defects generated on the surface of MWCNTs by the acid oxidation treatment. However, Pt nanoparticles are evenly deposited on the surface of PEDOT–MWCNTs with no agglomeration (Fig. 1.11e). The average particle size is 2.5 ± 0.3 nm, much smaller than that of the MWCNTs. In the case of Pt/MnO_x–PEDOT–MWCNTs (Fig. 1.11f), the average particle size of Pt nanoparticles is 2.5 ± 0.4 nm, which is almost

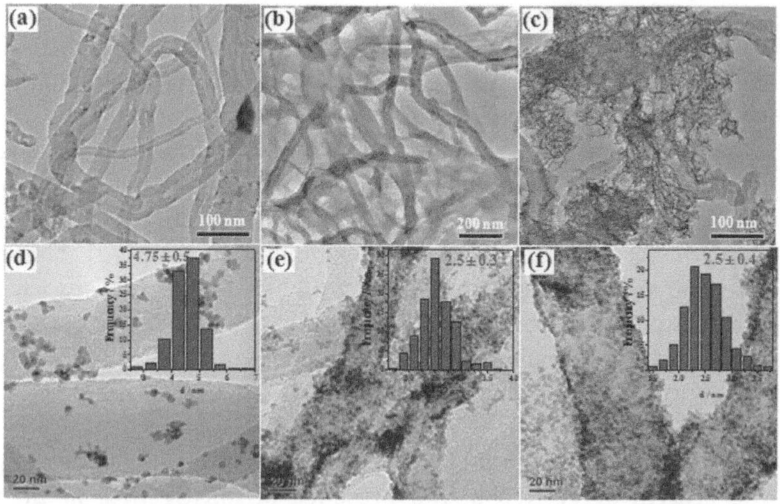

Figure 1.11 TEM images of functionalized MWCNTs (a), PEDOT–MWCNTs (b), and MnO_x–PEDOT–MWCNTs (c). TEM images of Pt/MWCNTs (d), Pt/PEDOT–MWCNTs (e), and Pt/MnO_x–PEDOT–MWCNTs (f) catalysts. The insets in d, e, and f are the corresponding size distribution histograms. Reprinted with permission from Wei et al. [167]. Copyright © 2013 Elsevier.

similar to that of the Pt/PEDOT–MWCNTs, indicating that the MnO_x cannot influence the size and dispersion of Pt nanoparticles. The electrochemical tests indicate that the methanol electro-oxidation activity and stability of the Pt/MnO_x–PEDOT–MWCNTs are significantly enhanced as compared with the Pt/PEDOT–MWCNT and Pt/MWCNT catalysts.

1.5 Control of Catalyst Surface Structures

Recent developments in nanoparticle (NP) synthesis have led to the formation of various monodisperse NPs with controlled size, shape, composition, and interparticle interactions [2, 169–171]. Because of their broad use in catalysis, photonics, electronics, plasmonics, optical sensing, biological labeling, imaging, and photothermal therapy, the controlled synthesis of noble-metallic nanocrystals with uniform geometries has been a research theme for decades.

Because the properties of noble–metallic nanocrystals are strongly determined by their size, shape, and surface structure, structural controlling of catalyst synthesis seems very important in obtaining highperforming catalysts.

As for metal NPs' shape or surface atomic arrangement, size and crystallinity, their high-index planes have a high density of low-coordinated surface atoms and generally exhibit much higher catalytic activity than those with low-index planes [172]. Therefore, the surface shape–controlled synthesis of NPs has attracted great attention, resulting in different shapes such as nanocubes [173, 174], nanorods [175, 176], nanonetwork structures [177], nanowires [178, 179], nanoflowers [180–182], and polyhedra [171, 183–187].

Gong et al. [175] found a new growth approach by investigating the growth of Pt NPs in the presence of Mn_3O_4 NPs. In their experiments, the pre-prepared Mn_3O_4 NPs were added to a mixture of oleyl amine and a Pt precursor, $Pt(acac)_2$. During heating at 180°C, $Pt(acac)_2$ was reduced by oleyl amine. In general, Mn_3O_4 NPs can act as seeds and the reduced Pt atoms can grow on the Mn_3O_4 NPs' surface to form core–shell or dumbbell NPs. However, unexpectedly, selective growth was not observed and newly formed Pt atoms nucleated separately. Interestingly, although seeding growth did not occur, the Mn_3O_4 NPs had a great impact on the growth of Pt NPs. More interestingly, the Pt NPs formed were rarely observed to be polycrystalline. Further growth centering the twinned polycrystalline Pt NPs could generate length-controlled one-dimensional Pt NRs and, importantly, it was found that high-index facets existed at the edge of the nanorods that were formed. All the Pt NRs grow from one center and exhibit match-like morphology (Fig. 1.12A,B). The thickness of NRs was about 5.6 nm (Fig. 1.12C). The HRTEM image in Fig. 1.12D gave the atomic packing of NRs obtained. Clearly, atomic defects and steps existed on the edge of the NRs, indicating the presence of high-index facets.

Tian et al. [183] made a breakthrough in synthesis of Pt nanocrystals enclosed with high-index facets by developing an electrochemical method. They used a two-step synthesis process. The first step is electrodeposition of polycrystalline Pt nanospheres ~750 nm in diameter on a glassy carbon (GC) electrode. After

Figure 1.12 The HRTEM images of Pt nanorods at different magnifications. Reproduced with permission from Gong et al. [175]. Copyright © 2011 Royal Society of Chemistry.

pre-deposition, a square-wave potential was applied to the Pt nanospheres in a solution containing 0.1 M H_2SO_4+ 30 mM ascorbic acid for 5∼60 min. The lower (E_L) and upper (E_U) potential limits were ca. −0.10 and 1.20 V vs. saturated calomel electrode (SCE), respectively, and $f = 10$ Hz. Under this condition, the Pt nanospheres are partially dissolved at the E_U, and provide low-concentrated Pt ions for the growth of new nanocrystals at the E_L. Interestingly, nearly all of the growing Pt nanocrystals on the GC surface are tetrahexahedral Pt nanocrystals (THH Pt NCs). The yield of the THH Pt NCs in the final product is >90%. Their average size (Heywood diameter) was 217 nm, with a standard deviation of 23 nm. It is found that the single-crystal THH NC is enclosed by 24 high-index facets such as {730}, {210}, and/or {520} surfaces that have a large density of atomic steps and dangling bonds. These high-energy surfaces were stable thermally (to 800°C) and chemically, and exhibited a much enhanced (up to 400%) catalytic activity for

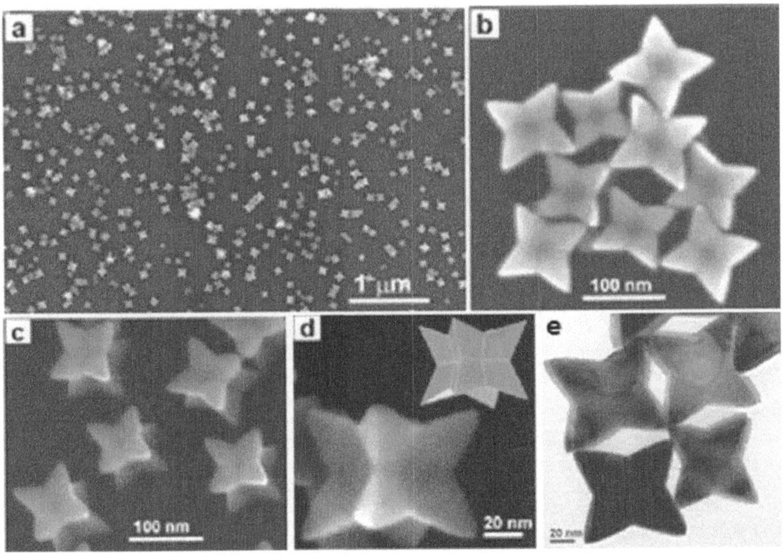

Figure 1.13 (a–c) Typical (a) large-area, (b) enlarged, and (c) tilted SEM images of the as-prepared concave Pt nanocrystals. (d) High-magnification SEM image of a single concave Pt nanocrystal. The top-right inset shows an ideal geometrical model of the concave Pt nanocrystal with the same orientation as the nanocrystal in the SEM image. (e) TEM images of the concave Pt nanocrystals. Reprinted with permission from Huang et al. [185]. Copyright © 2011 American Chemical Society.

equivalent Pt surface areas for the electro-oxidation of small organic fuels such as formic acid and ethanol.

Huang et al. [185] prepared concave Pt nanocrystals enclosed by both high-index {411} and {100} facets via a facile wet-chemical route, as shown in Fig. 1.13. The use of amine was demonstrated to be most essential for the formation of the concave nanocrystals. Because of the presence of the high-index {411} exposed facets, the as-prepared concave Pt nanocrystals exhibited a higher electrocatalytic activity per unit surface area than either commercial Pt black or Pt/C in the electro-oxidation of both formic acid and ethanol. Further studies are still required in order to understand fully how small molecules such as amines determine the formation of high-index Pt facets. The as-prepared Pt nanocrystals were first characterized by scanning electron microscopy (SEM)

(Fig. 1.13a). The product consisted of uniform four-armed starlike particles with a yield above 90%. The apex angle of each arm was measured to be 53°. In the SEM image at a higher magnification, the four arms are much brighter than the center region of the starlike particles (Fig. 1.13b), indicating the possible presence of a concave feature in the particles. The as-prepared concave particles are essentially octapods having eight trigonal pyramidal arms (Fig. 1.13c). In the SEM image before tilting, most of the particles were observed as four-armed stars simply because they prefer to lie flat on the substrate using four of their eight arms. When hindered by neighboring particles, some octapods were able to settle on the substrate using only two arms (Fig. 1.13d), allowing direct three-dimensional visualization of the concave particles. At the core of the octapod, basins surrounded by every four trigonal-pyramidal arms can be observed. On the basis of these observations, we have proposed a structural model (inset in Fig. 1.13d) for as-prepared Pt nanocrystals.

Wei et al. [187] developed the electrochemically shape-controlled synthesis in deep eutectic solvents (DESs) for the preparation of Pt nanocrystals enclosed by high-index facets. Monodispersed concave tetrahexahedral Pt nanocrystals (THH Pt NCs) were prepared through this new route. The as-prepared concave THH Pt NCs were carefully characterized by SEM and TEM (Fig. 1.14). It can be clearly seen that the Pt NCs appear to have a concave cubic morphology, and the faces of each cube have contrast lines in the shape of an "X" (Fig. 1.14a,b). The SEM images (Fig. 1.14c) of concave THH Pt NCs tilted 0°, 45°, and 90° illustrated that each concave face of the Pt NCs has four trigonal facets. The as-prepared concave Pt NCs were bounded with {910} and vicinal high-index facets, which exhibited superior catalytic activity and stability to those of the commercial Pt black catalyst for ethanol electro-oxidation (Fig. 1.14d). They also demonstrated that the electrochemically shape-controlled synthesis in DESs could prove advantageous in controlling the size and shape of Pt NCs without the addition of seeds, surfactants, or other chemicals, and could be applied in the synthesis of other noble metal NCs with high surface energy and high catalytic activity.

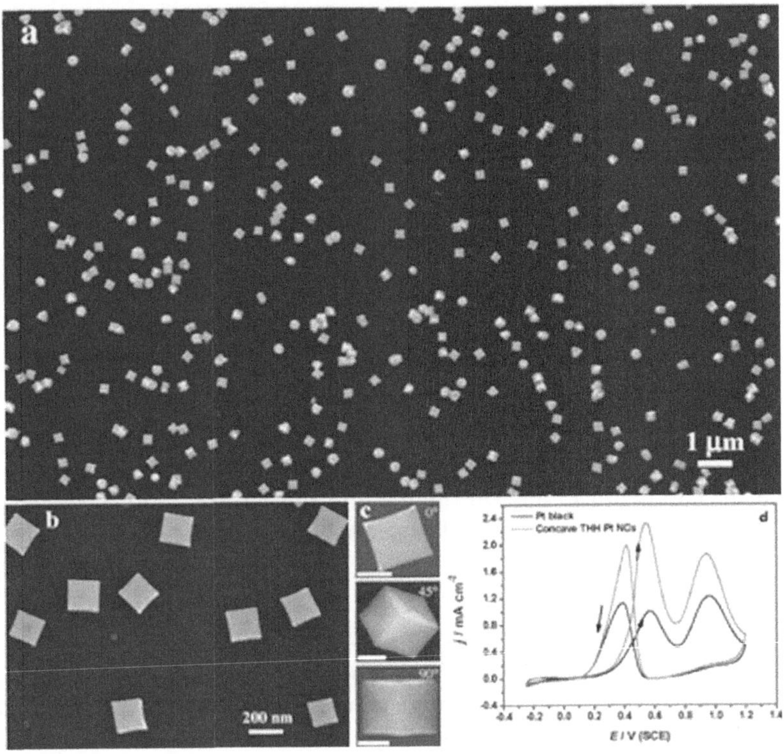

Figure 1.14 (a) Large-area, (b) enlarged SEM images of concave THH Pt NCs electrodeposited on GC in 19.3 mM H_2PtCl_6/DESs solution at 80°C by a programmed electrodeposition method: $E_L = -1.3$ V, $E_U = -0.3$ V, growth time = 60 min. (c) SEM images of concave THH Pt NCs tilted 0°, 45°and 90°to illustrate the concave faces. Scale bar of the inset is 100 nm. (d) Cyclic voltammmograms (50 mV s^{-1}) of ethanol oxidation on concave THH Pt NCs. Reprinted with permission from Wei et al. [187]. Copyright © 2012 American Chemical Society.

1.6 Fundamental Understanding of Catalystic for New Catalyst Design

A more fundamental understanding is essential in designing alternative and innovative Pt-free catalysts for practical applications of fuel cells such as DMFCs. DMFCs are considered to be a promising commercial energy conversion device for portable electronics with many advantages. As is known, the catalyst coated membrane technique could achieve high catalyst utilization because of better contact between the membrane and the catalyst layer [188]. However, the cracks, irregular pore structure, and agglomeration of catalyst particles in catalyst layers, particularly the anode catalyst layer, can decrease the performance of a DMFC. To improve the performance of the DMFC, previous studies mainly focused on developing new catalysts with high activity [3, 44, 45, 189], new proton exchange membranes [190], and modification of the Nafion® membrane [191]. In fact, the morphology and structure of the anode catalyst layers can also significantly influence the fuel cell performance. To optimize the catalyst/catalyst layer performance, theoretical modeling is very useful to achieve fundamental understanding of the factors affecting fuel cell performance. The results obtained can be used to guide the catalyst downs-election and new catalyst design.

A theoretical model for catalyst layer (CL) performance could also be used to determine the optimal electrolyte content and catalyst loading in the CL, as well as the current density [192]. Antolini [193] proposed an empirical model to evaluate the contribution of alloyed and non-alloyed tin in Pt_z–Sn/C catalysts to the performance of direct ethanol fuel cells (DEFCs). The model was based on the hypotheses that (i) all Pt is present in the forms of a $Pt_{(1-x)}Sn_x$ solid solution, and (ii) the effect on the EOR of the interactions between $Pt_{(1-x)}Sn_x$ and SnO_2 is the same as that between Pt and SnO_2. The model predicted the performance dependence of a single DEFC on the total Sn content and the degree of alloying of Pt_z–Sn/C catalysts used as the anode material.

1.7 Challenges and Possible Research Directions

Great progress has been achieved during the last decades in fuel cell science and technology, especially in some application areas such as portable, transportation, and stationary power sources [194, 195]. However, the commercialization of DAFCs is still hindered by some challenges. Two major ones are insufficient catalytic performance (both low catalytic activity and stability) and high cost. Regarding the insufficient performance of catalysts, the major factors are the sluggish alcohol oxidation kinetics even on some state-of-the-art anode catalysts, and the instability of the catalysts, as well as alcohol crossover through the membrane, which not only depresses cathode performance but also reduces fuel efficiency. The current practical anode catalysts for DAFCs are Pt-based catalysts, which have several disadvantages: (i) Pt is an expensive and scarce metal. (ii) Pt catalysts are readily poisoned by CO and other contaminants during the oxidation of fuel. Although some Pt-based alloy catalysts exhibit enhanced tolerance of CO, Pt still remains a major limiting factor in the overall performance. (iii) Pt catalysts are not stable in the acidic fuel cell reaction conditions, because they tend to dissolute and coalesce. This reduces their active surface area, deteriorates their performance, and limits their lifetime. Currently, most research in anode catalyst development is focused on exploring new stable anode catalysts, catalyst supports, as well as their optimal interaction to effectively enhance the alcohol electro-oxidation kinetics, and seeking new membranes that have a low alcohol crossover. In an effort to reduce cathode performance depression induced by alcohol crossover, some activities on alcohol tolerant cathode catalysts have also been carried out.

As regards the high cost of the DAFCs, the major contributor is the expensive Pt-based catalysts. Pt-based DAFC anode catalyst loadings must drop to a level of <1.0 mg cm^{-2} from the present 2.0–8.0 mg cm^{-2}, depending on applications. Loading reduction by increasing Pt utilization is one of the research directions. Alloying and nanoparticle-supporting strategies could dramatically reduce the Pt content in the catalysts without performance compromise. Non-noble catalyst development is the other approach for catalyst cost reduction.

1.8 Summary

In this chapter, several types of important anode catalysts for DAFCs were reviewed with a detailed discussion of their catalytic activity, stability, synthesis, characterization, as well as performance validation. The interaction between the catalyst metal particles and their support materials was emphasized in terms of its effects on the catalytic activity and stability. The important role of catalyst support materials in overall catalyst performance was also emphasized in the chapter. To give the reader a complete picture of the anode catalysts and their associated research and development in the last several decades, the current state of technology, the progress in research and development, and both technical and commercialization challenges were also described and discussed. To facilitate the research and development in this topic, several research directions to overcome the challenges were also proposed in this chapter.

References

1. Sharma, S., and Pollet, B. G. (2012). Support materials for PEMFC and DMFC electrocatalysts: a review, *J. Power Sources*, **208**, 96–119.
2. Chen, A. C., and Holt-Hindle, P. (2010). Platinum-based nanostructured materials: synthesis, properties, and applications, *Chem. Rev.*, **110**, 3767–3804.
3. Liu, H. S., Song, C. J., Zhang, L., Zhang, J. J., Wang, H. J., and Wilkinson, D. P. (2006). A review of anode catalysis in the direct methanol fuel cell, *J. Power Sources*, **155**, 95–110.
4. Antolini, E. (2007). Catalysts for direct ethanol fuel cells, *J. Power Sources*, **170**, 1–12.
5. Song, S., Liang, Y., Li, Z., Wang, Y., Fu, R., Wu, D., and Tsiakaras, P. (2010). Effect of pore morphology of mesoporous carbons on the electrocatalytic activity of Pt nanoparticles for fuel cell reactions, *Appl. Catal. B*, **98**, 132–137.
6. Li, H., Sun, G., Cao, L., Jiang, L., and Xin, Q. (2007). Comparison of different promotion effect of PtRu/C and PtSn/C electrocatalysts for ethanol electro-oxidation, *Electrochim. Acta*, **52**, 6622–6629.
7. Coutanceau, C., Brimaud, S., Lamy, C., Léger, J. M., Dubau, L., Rousseau, S., and Vigier F. (2008). Review of different methods for developing nanoelectrocatalysts for the oxidation of organic compounds, *Electrochim. Acta*, **53**, 6865–6880.
8. An, X. S., Fan, Y. J., Chen, D. J., Wang, Q., Zhou, Z. Y., and Sun, S. G. (2011). Enhanced activity of rare earth doped PtRu/C catalysts for methanol electro-oxidation, *Electrochim. Acta*, **56**, 8912–8918.

9. Wang, Z. B., Zuo, P. J., Wang, G. J., Du, C. Y., and Yin, G. P. (2008). Effect of Ni on PtRu/C catalyst performance for ethanol electrooxidation in acidic medium, *J. Phys. Chem. C*, **112**, 6582–6587.
10. Sahin, O., and Kivrak, H. (2013). A comparative study of electrochemical methods on Pt–Ru DMFC anode catalysts: the effect of Ru addition, *Int. J. Hydrogen Energy*, **38**, 901–909.
11. Lobato, J., Canizares, P., Ubeda, D., Pinar, F. J., and Rodrigo, M. A. (2011). Testing PtRu/CNF catalysts for a high temperature polybenzimidazole-based direct ethanol fuel cell: effect of metal content, *Appl. Catal. B*, **106**, 174–180.
12. Corpuz, A. R., Olson, T. S., Joghee, P., Pylypenko, S., Dameron, A. A., Dinh, H. N., O'Neill, K. J., Hurst, K. E., Bender, G., Gennett, T., Pivovar, B. S., Richards, R. M., and O'Hayre, R. P. (2012). Effect of a nitrogen-doped PtRu/carbon anode catalyst on the durability of a direct methanol fuel cell, *J. Power Sources*, **217**, 142–151.
13. Zhiani, M., Jalili, J., Rezaei, B., and Taghiabadi, M. M. (2013). Methanol electrooxidation on synthesized PtRu nanocatalyst supported on acetylene black in half cell and in direct methanol fuel cell, *Int. J. Hydrogen Energy*, **38**, 5419–5424.
14. Zhao, H. B., Yang, J., Li, L., Li, H., Wang, J. L., and Zhang, Y. M. (2009). Effect of over-oxidation treatment of Pt–Co/polypyrrole–carbon nanotube catalysts on methanol oxidation, *Int. J. Hydrogen Energy*, **34**, 3908–3914.
15. Antolini, E., and Gonzalez, E. R. (2010). A simple model to assess the contribution of alloyed and non-alloyed platinum and tin to the ethanol oxidation reaction on Pt–Sn/C catalysts: application to direct ethanol fuel cell performance, *Electrochim. Acta*, **55**, 6485–6490.
16. Zignani, S. C., Baglio, V., Linares, J. J., Monforte, G., Gonzalez, E. R., and Arico, A. S. (2013). Endurance study of a solid polymer electrolyte direct ethanol fuel cell based on a Pt–Sn anode catalyst, *Int. J. Hydrogen Energy*, **38**, 11576–11582.
17. Purgato, F. L. S., Pronier, S., Olivi, P., Andrade, A. R., de. Léger, J. M., Tremiliosi-Filho, G., and Kokoh, K. B. (2012). Direct ethanol fuel cell: electrochemical performance at 90°C on Pt and PtSn/C electrocatalysts, *J. Power Sources*, **198**, 95–99.
18. Garcia-Rodriguez, S., Somodi, F., Borbath, I., Margitfalvi, J. L., Pena, M. A., Jose Fierro, L. G., and Rojas, S. (2009). Controlled synthesis of Pt–Sn/C fuel cell catalysts with exclusive Sn–Pt interaction application in CO and ethanol electrooxidation reactions, *Appl. Catal. B*, **91**, 83–91.

19. Li, L. R., Huang, M. H., Liu, J. J., and Guo, Y. L. (2011). PtxSn/C electrocatalysts synthesized by improved microemulsion method and their catalytic activity for ethanol oxidation, *J. Power Sources*, **196**, 1090–1096.
20. Yang, M. L. (2013). Catalytic activities of PtBi nanoparticles toward methanol electrooxidation in acid and alkaline media, *J. Power Sources*, **229**, 42–47.
21. Teliz, E., Díaz, V., Pérez, I., Corengia, M., and Zinola, C. F. (2012). Carbon supported Pt, Ru and Mo catalysts for methanol electrooxidation, *Int. J. Hydrogen Energy*, **37**, 14761–14768.
22. Herrera-Méndez, H. D., Roquero, P., Smit, M. A., and Ordóñez, L. C. (2011). Carbon-supported platinum molybdenum electro-catalysts and their electro-activity toward ethanol oxidation, *J. Electrochem. Sci.*, **6**, 4454–4469.
23. Yang, X., Yang, Q. D., Xu, J., and Lee, C. S. (2012). Bimetallic PtPd nanoparticles on Nafion-graphene film as catalyst for ethanol electro-oxidation, *J. Mater. Chem.*, **22**, 8057–8062.
24. Seweryn, J., and Lewera, A. (2012). Electrooxidation of ethanol on carbon-supported Pt-Pd nanoparticles, *J. Power Sources*, **205**, 264–271.
25. Wang, Y. X., Zhou, H. J., Sun, P. C., and Chen, T. H. (2014). Exceptional methanol electro-oxidation activity by bimetallic concave and dendritic Pt-Cu nanocrystals catalysts, *J. Power Sources*, **245**, 663–670.
26. Habibi, B., and Dadashpour, E. (2013). Electrooxidation of 2-propanol and 2-butanol on the Pt-Ni alloy nanoparticles in acidic media, *Electrochim. Acta*, **88**, 157–164.
27. He, H. C., Xiao, P., Zhou, M., Liu, F., Yu, S. J., Qiao, L., and Zhang, Y. H. (2013). PtNi alloy nanoparticles supported on carbon-doped TiO2 nanotube arrays for photo-assisted methanol oxidation, *Electrochim. Acta*, **88**, 782–789.
28. Wang, Q., Sun, G. Q., Jiang, L. H., Xin, Q., Sun, S. G., Jiang, Y. X., Chen, S. P., Jusys, Z., and Behm, R. J. (2007). Adsorption and oxidation of ethanol on colloid-based Pt/C, PtRu/C and Pt3Sn/C catalysts: in situ FTIR spectroscopy and on-line DEMS studies, *Phys. Chem. Chem. Phys.*, **9**, 2686–2696.
29. Wang, H., Jusys, Z., and Behm, R. J. (2006). Ethanol electro-oxidation on carbon-supported Pt, PtRu and Pt3Sn catalysts: a quantitative DEMS study, *J. Power Sources*, **154**, 351–359.

30. Tayal, J., Rawat, B., and Basu, S. (2011). Bi-metallic and tri-metallic Pt-Sn/C, Pt-Ir/C, Pt-Ir-Sn/C catalysts for electro-oxidation of ethanol in direct ethanol fuel cell, *Int. J. Hydrogen Energy*, **36**, 14884–14897.
31. Lee, E., Murthy, A., and Manthiram, A. (2011). Effect of Mo addition on the electrocatalytic activity of Pt–Sn–Mo/C for direct ethanol fuel cells, *Electrochim. Acta*, **56**, 1611–1618.
32. Chu, Y. H., and Shul, Y. G. (2010). Combinatorial investigation of Pt–Ru–Sn alloys as an anode electrocatalysts for direct alcohol fuel cells, *Int. J. Hydrogen Energy*, **35**, 11261–11270.
33. Tayal, J., Rawat, B., and Basu, S. (2012). Effect of addition of rhenium to Pt-based anode catalysts in electro-oxidation of ethanol in direct ethanol PEM fuel cell, *Int. J. Hydrogen Energy*, **37**, 4597–4605.
34. Almeida, T. S., Kokoh, K. B., and De Andrade, A. R. (2011). Effect of Ni on Pt/C and PtSn/C prepared by the Pechini method, *Int. J. Hydrogen Energy*, **36**, 3803–3810.
35. Bonesi, A. R., Moreno, M. S., Triaca, W. E., and Castro Luna, A. M. (2010). Modified catalytic materials for ethanol oxidation, *Int. J. Hydrogen Energy*, **35**, 5999–6004.
36. Beyhan, S., Leger, J. M., and Kadrgan, F. (2013). Pronounced synergetic effect of the nano-sized PtSnNi/C catalyst for ethanol oxidation in direct ethanol fuel cell, *Appl. Catal. B*, **130–131**, 305–313.
37. De Souza, R. F. B., Parreira, L. S., Silva, J. C. M., Simoes, F. C., Calegaro, M. L., Giz, M. J., Camara, G. A., Neto, A. O., and Santos, M. C. (2011). PtSnCe/C electrocatalysts for ethanol oxidation: DEFC and FTIR "in situ" studies, *Int. J. Hydrogen Energy*, **36**, 11519–11527.
38. Nakagawa, N., Kaneda, Y., Wagatsuma, M., and Tsujiguchi, T. (2012). Product distribution and the reaction kinetics at the anode of direct ethanol fuel cell with Pt/C, PtRu/C and PtRuRh/C, *J. Power Sources*, **199**, 103–109.
39. Tsiouvaras, N., Martınez-Huerta, M. V., Paschos, O., Stimming, U., Fierro, J. L. G., and Pena, M. A. (2010). PtRuMo/C catalysts for direct methanol fuel cells: effect of the pretreatment on the structural characteristics and methanol electrooxidation, *Int. J. Hydrogen Energy*, **35**, 11478–11488.
40. Chang, Y. W., Liu, C. W., Wei, Y. C., and Wang, K. W. (2009). Promotion of PtRu/C anode catalysts for ethanol oxidation reaction by addition of Sn modifier, *Electrochem. Commun.*, **11**, 2161–2164.
41. Ammam, M., Prest, L. E., Pauric, A. D., and Easton, E. B. (2012). Synthesis, characterization and catalytic activity of binary PtMn/C

alloy catalysts towards ethanol oxidation, *J. Electrochem. Soc.*, **159**, B195–B200.

42. Ammam, M., and Easton, E. B. (2012). Ternary PtMnX/C (X = Fe, Co, Ni, Cu, Mo and, Sn) alloy catalysts for ethanol electrooxidation, *J. Electrochem. Soc.*, **159**, B635–B640.
43. Ammam, M., and Easton, E. B. (2012). Quaternary PtMnCuX/C (X = Fe, Co, Ni, and Sn) and PtMnMoX/C (X = Fe, Co, Ni, Cu and Sn) alloys catalysts: synthesis, characterization and activity towards ethanol electrooxidation, *J. Power Sources*, **215**, 188–198.
44. Zhou, C. M., Wang, H. J., Peng, F., Liang, J. H., Yu, H., and Yang, J. (2009). MnO2/CNT supported Pt and PtRu nanocatalysts for direct methanol fuel cells, *Langmuir*, **25**, 7711–7717.
45. Huang, H. J., Chen, Q., He, M., Sun, X., and Wang, X. (2013). A ternary Pt/MnO2/graphene nanohybrid with an ultrahigh electrocatalytic activity toward methanol oxidation, *J. Power Sources*, **239**, 189–195.
46. Peng, F., Zhou, C. M., Wang, H. J., Yu, H., Liang, J. H., and Yang, J. (2009). The role of RuO2 in the electrocatalytic oxidation of methanol for direct methanol fuel cell, *Catal. Commun.*, **10**, 533–537.
47. Chu, Y. Y., Wang, Z. B., Jiang, Z. Z., Gu, D. M., and Yin, G. P. (2011). A novel structural design of a Pt/C–CeO2 catalyst with improved performance for methanol electro-oxidation by β-cyclodextrin carbonization, *Adv. Mater.*, **23**, 3100–3104.
48. Ou, D. R., Mori, T., Togasaki, H., Takahashi, M., Ye, F., and Drennan, J. (2011). Microstructural and metal–support interactions of the Pt–CeO2/C catalysts for direct methanol fuel cell application, *Langmuir*, **27**, 3859–3866.
49. You, E., Guzman–Blas, R., Nicolau, E., Scibioh, M. A., Karanikas, C. F., Watkins, J. J., and Cabrera, C. R. (2012). Co–deposition of Pt and ceria anode catalyst in supercritical carbon dioxide for direct methanol fuel cell applications, *Electrochim. Acta*, **75**, 191–200.
50. Guillen–Villafuerte, O., Garcıa, G., Rodrıguez, J. L., Pastor, E., Guil–Lopez, R., Nieto, E., and Fierro, J. L. G. (2013). Preliminary studies of the electrochemical performance of Pt/X@MoO3/C (X = Mo2C, MoO2, Mo°) catalysts for the anode of a DMFC: influence of the Pt loading and Mo–phase, *Int. J. Hydrogen Energy*, **38**, 7811–7821.
51. Baglio, V., Sebastián, D., D'Urso, C., Stassi, A., Amin, R. S., El–Khatib, K. M., and Aricò, A. S. (2014). Composite anode electrode based on iridium oxide promoter for direct methanol fuel cells, *Electrochim. Acta*, **128**, 304–310.

52. Guo, D. J. (2011). Electrooxidation of ethanol on novel multi-walled carbon nanotube supported platinum–antimony tin oxide nanoparticle catalysts, *J. Power Sources*, **196**, 679–682.
53. Yu, L. H., and Xi, J. Y. (2012). TiO2 nanoparticles promoted Pt/C catalyst for ethanol electro-oxidation, *Electrochim. Acta*, **67**, 166–171.
54. Saida, T., Ogiwara, N., Takasu, Y., and Sugimoto, W. (2010). Titanium oxide nanosheet modified PtRu/C electrocatalyst for direct methanol fuel cell anodes, *J. Phys. Chem. C*, **114**, 13390–13396.
55. Yang, C., Van der Laak, N., Chan, K., and Zhang, X. (2012). Microwave-assisted microemulsion synthesis of carbon supported Pt–WO3 nanoparticles as an electrocatalyst for methanol oxidation, *Electrochim. Acta*, **75**, 262–272.
56. Yan, Z., Xie, J., Fang, Y., Chen, M., Wei, X., and Lü, X. (2014). WO3 nanoparticles synthesized through ion exchange route as Pt electrocatalyst support for alcohol oxidation, *Fuel Cells*, **14**, 2291–2295.
57. Xu, C. W., Tian, Z. Q., Shen, P. K., and Jiang S. P. (2008). Oxide (CeO2, NiO, Co3O4 and Mn3O4)-promoted Pd/C electrocatalysts for alcohol electrooxidation in alkaline media, *Electrochim. Acta*, **53**, 2610–2618.
58. Hu, F. P., Wang, Z., Li, Y., Li, C., Zhang, X., and Shen, P. K. (2008). Improved performance of Pd electrocatalyst supported on ultrahigh surface area hollow carbon spheres for direct alcohol fuel cells, *J. Power Sources*, **177**, 61–66.
59. Yi, Q. F., Niu, F. J., and Sun, L. Z. (2011). Fabrication of novel porous Pd particles and their electroactivity towards ethanol oxidation in alkaline media, *Fuel*, **90**, 2617–2623.
60. Wei, W. T., and Chen, W. (2012). "Naked" Pd nanoparticles supported on carbon nanodots as efficient anode catalysts for methanol oxidation in alkaline fuel cells, *J. Power Sources*, **204**, 85–88.
61. Yan, Z., Zhang, M., Xie, J., and Shen, P. K. (2013). Vanadium carbide and graphite promoted Pd electrocatalyst for ethanol oxidation in alkaline media, *J. Power Sources*, **243**, 336–342.
62. Shen, P. K., and Xu, C. W. (2006). Alcohol oxidation on nanocrystalline oxide Pd/C promoted electrocatalysts, *Electrochem. Commun.*, **8**, 184–188.
63. Bianchini, C., and Shen, P. K. (2009). Palladium-based electrocatalysts for alcohol oxidation in half cells and in direct alcohol fuel cells, *Chem. Rev.*, **109**, 4183–4206.

64. Sawangphruk, M., Krittayavathananon, A., and Chinwipas, N. (2013). Ultraporous palladium on flexible graphene–coated carbon fiber paper as high-performance electro-catalysts for the electro-oxidation of ethanol, *J. Mater. Chem. A*, **1**, 1030–1034.

65. Wang, X. G., Wang, W. M., Qi, Z., Zhao, C. C., Ji, H., and Zhang, Z. H. (2009). High catalytic activity of ultrafine nanoporous palladium for electro-oxidation of methanol, ethanol, and formic acid, *Electrochem. Commun.*, **11**, 1896–1899.

66. Shen, S. Y., Zhao, T. S., Xu, J. B., and Li, Y. S. (2010). Synthesis of PdNi catalysts for the oxidation of ethanol in alkaline direct ethanol fuel cells, *J. Power Sources*, **195**, 1001–1006.

67. Antolini, E. (2009). Palladium in fuel cell catalysis, *Energy Environ. Sci.*, **2**, 915–931.

68. Ma, L., He, H., Hsu, A., and Chen, R. (2013). PdRu/C catalysts for ethanol oxidation in anion–exchange membrane direct ethanol fuel cells, *J. Power Sources*, **241**, 696–702.

69. Wang, Y., Zhao, Y., Yin, J., Liu, M., Dong, Q., and Su, Y. (2014). Synthesis and electrocatalytic alcohol oxidation performance of Pd–Co bimetallic nanoparticles supported on graphene, *Int. J. Hydrogen Energy*, **39**, 1325–1335.

70. Dutta, A., and Datta, J. (2012). Outstanding catalyst performance of PdAuNi nanoparticles for the anodic reaction in an alkaline direct ethanol (with anion–exchange membrane) fuel cell, *J. Phys. Chem. C*, **116**, 25677–25688.

71. Wang, Y., Shi, F. F., Yang, Y. Y., and Cai, W. B. (2013). Carbon supported Pd–Ni–P nanoalloy as an efficient catalyst for ethanol electro-oxidation in alkaline media, *J. Power Sources*, **243**, 369–373.

72. Jia, F. L., Wong, K. W., and Du, R. X. (2009). Direct growth of highly catalytic palladium nanoplates array onto gold substrate by a template–free electrochemical route, *Electrochem. Commun.*, **11**, 519–521.

73. Jia, F. L., Wong, K. W., and Zhang, L. Z. (2009). Electrochemical synthesis of nanostructured palladium of different morphology directly on gold substrate through a cyclic deposition/dissolution route, *J. Phys. Chem. C*, **113**, 7200–7206.

74. Wang, X. G., Wang, W. M., Qi, Z., Zhao, C. C., Ji, H., and Zhang, Z. H. (2010). Electrochemical catalytic activities of nanoporous palladium rods for methanol electro-oxidation, *J. Power Sources*, **195**, 6740–6747.

75. Yin, Z., Zheng, H. J., Ma, D., and Bao, X. H. (2009). Porous palladium nanoflowers that have enhanced methanol electro-oxidation activity, *J. Phys. Chem. C*, **113**, 1001–1005.

76. Hu, F. P., and Chen, W. X. (2011). End-opened carbon nanotube array supported Pd as anode for alkaline fuel cell, *Electrochem. Commun.*, **13**, 955–958.

77. Liu, Z., Zhao, B., Guo, C., Sun, Y., Shi, Y., Yang, H., and Li, Z. (2010). Carbon nanotube/raspberry hollow Pd nanosphere hybrids for methanol, ethanol, and formic acid electro-oxidation in alkaline media, *J. Colloid Interf. Sci.*, **351**, 233–238.

78. Tan, Q., Du, C. Y., Yin, G. P., Zuo, P. J., Cheng, X. Q., and Chen, M. (2012). Highly efficient and stable nonplatinum anode catalyst with Au@Pd core–shell nanostructures for methanol electrooxidation, *J. Catal.*, **295**, 217–222.

79. Kakati, N., Maiti, J., Lee, S. H., and Yoon, Y. S. (2012). Core shell like behavior of PdMo nanoparticles on multiwall carbon nanotubes and their methanol oxidation activity in alkaline medium, *Int. J. Hydrogen Energy*, **37**, 19055–19064.

80. Cao, L., Sun, G. Q., Li, H. Q., and Xin, Q. (2007). Carbon-supported IrSn catalysts for a direct ethanol fuel cell, *Electrochem. Commun.*, **9**, 2541–2546.

81. Uribe-Godínez, J., García-Montalvo, V., and Jiménez-Sandoval, O. (2014). A novel Rh–Ir electrocatalyst for the oxygen reduction reaction and the hydrogen and methanol oxidation reactions, *Int. J. Hydrogen Energy*, **39**, 9121–9127.

82. Zhang, H., Gu, C. D., Huang, M. L., Wang, X. L., and Tu, J. P. (2013). Anchoring three-dimensional network structured Ni–P nanowires on reduced graphene oxide and their enhanced electrocatalytic activity towards methanol oxidation, *Electrochem. Commun.*, **35**, 108–111.

83. Abdel Hameed, R. M., and El-Khatib, K. M. (2010). Ni–P and Ni–Cu–P modified carbon catalysts for methanol electro-oxidation in KOH solution, *Int. J. Hydrogen Energy*, **35**, 2517–2529.

84. Raoof, J. B., Ojani, R., and Hosseini, S. R. (2011). A novel, effective and low cost catalyst for methanol oxidation based on nickel ions dispersed onto poly(o-toluidine)/Triton X-100 film at the surface of multi-walled carbon nanotube paste electrode, *J. Power Sources*, **196**, 1855–1863.

85. Zhang, R., Zhang, W., Gao, L. Z., Zhang, J., Li, P., Wang, W. Y., and Li, R. F. (2013). Electrocatalytic oxidation of methanol on Ni(II)Salen complex encapsulated with type-Y zeolite, *Appl. Catal. A*, **466**, 264–271.

86. Oh, Y., Kim, S. K., Peck, D. H. Jang, J. S., Kim, J., and Jung, D. H. (2014). Improved performance using tungsten carbide/carbon nanofiber based anode catalysts for alkaline direct ethanol fuel cells, *Int. J. Hydrogen Energy*, **39**(28), 15907–15912.

87. Antolini, E. (2009). Carbon supports for low-temperature fuel cell catalysts, *Appl. Catal. B*, **88**, 1–24.

88. Calvillo, L., Celorrio, V., Moliner, R., Garcia, A. B., Caméan, I., and Lazaro, M. J. (2013). Comparative study of Pt catalysts supported on different high conductive carbon materials for methanol and ethanol oxidation, *Electrochim. Acta*, **102**, 19–27.

89. Chen, Y., Zhang, G., Ma, J., Zhou, Y., Tang, Y., and Lu, T. (2010). Electro-oxidation of methanol at the different carbon materials supported Pt nano-particles, *Int. J. Hydrogen Energy*, **35**, 10109–10117.

90. Gattia, D. M., Antisari, M. V., Giorgi, L., Marazzi, R., Piscopiello, E., Montone, A., Bellitto, S., Licoccia, S., and Traversa, E. (2009). Study of different nanostructured carbon supports for fuel cell catalysts, *J. Power Sources*, **194**, 243–251.

91. Machado, B. F., Marchionni, A., Bacsa, R. R., Bellini, M., Beausoleil, J., Oberhauser, W., Vizza, F., and Serp, P. (2013). Synergistic effect between few layer graphene and carbon nanotube supports for palladium catalyzing electrochemical oxidation of alcohols, *J. Energy Chem.*, **22**, 296–304.

92. Hong, W., Liu, Y., Wang, J., and Wang, E. (2013). A new kind of highly active hollow flower-like NiPdPt nanoparticles supported by multiwalled-carbon nanotubes toward ethanol electrooxidation, *J. Power Sources*, **241**, 751–755.

93. Yoo, E., Okada, T., Kizuka, T., and Nakamura, J. (2008). Effect of carbon substrate materials as a Pt–Ru catalyst support on the performance of direct methanol fuel cells, *J. Power Sources*, **180**, 221–226.

94. Saito, R., Fujita, M., Dresselhaus, G., and Dresselhaus, M. (1992). Electronic structure of chiral graphene tubules, *Appl. Phys. Lett.*, **60**, 2204–2206.

95. Kasumov, A. Y., Bouchiat, H., Reulet, B., Stephan, O., Khodos, I. I., Gorbatov, Y. B., and Colliex, C. (1998). Conductivity and atomic structure of isolated multiwalled carbon nanotubes, *Europhys. Lett.*, **43**, 89–94.

96. Yaldagard, M., Jahanshahi, M., and Seghatoleslami, N. (2013). Carbonaceous nanostructured support materials for low temperature fuel cell electrocatalysts—a review, World, *J. Nano Sci. Eng.*, **3**, 121–153.

97. Iijima, S. (1991). Helical microtubules of graphitic carbon, *Nature*, **354**, 56–58.
98. Ebbesen, T. W., Hiura, H., Bisher, M. E., Treacy, M. M. J., Shreeve-Keyer, J. L., and Haushalter, R. C. (1996). Decoration of carbon nanotubes, *Adv. Mater.*, **8**, 155–157.
99. Huang, M., Dong, G., Wu, C., and Guan, L. (2014). Preparation of highly dispersed Pt nanoparticles supported on single-walled carbon nanotubes by a microwave-assisted polyol method and their remarkably catalytic activity, *Int. J. Hydrogen Energy*, **39**, 4266–4273.
100. Trogadas, P., Fuller, T. F., and Strasser, P. (2014). Carbon as catalyst and support for electrochemical energy conversion, *Carbon*, **75**, 5–42.
101. Hoa, L. Q., Vestergaard, M. C., Yoshikawa, H., Saito, M., and Tamiya, E. (2011). Functionalized multi-walled carbon nanotubes as supporting matrix for enhanced ethanol oxidation on Pt-based catalysts, *Electrochem. Commun.*, **13**, 746–749.
102. Kuang, Y. J., Cui, Y., Zhang, Y. S., Yu, Y. M., Zhang, X. H., and Chen, J. H. (2012). A strategy for the high dispersion of PtRu nanoparticles onto carbon nanotubes and their electrocatalytic oxidation of methanol, *Chem. Eur. J.*, **18**, 1522–1527.
103. Singh, R. N., Singh, A., and Anindita. (2009). Electrocatalytic activity of binary and ternary composite films of Pd, MWCNT, and Ni for ethanol electro-oxidation in alkaline solutions, *Carbon*, **47**, 271–278.
104. Li, S. W., Dong, Z. P., Yang, H. L., Guo, S. J., Gou, G. L., Ren, R., Zhu, Z. J., Jin, J., and Ma, J. T. (2013). Microenvironment effects in electrocatalysis: ionic-liquid-like coating on carbon nanotubes enhances the Pd-electrocatalytic alcohol oxidation, *Chem. Eur. J.*, **19**, 2384–2391.
105. Yang, X., Liu, X., Meng, X., Wang, X., Li, G., Shu, C., Jiang, L., and Wang, C. (2013). Self-assembly of highly dispersed Pt and PtRu nanoparticles on perylene diimide derivatives functionalized carbon nanotubes as enhanced catalysts for methanol electro-oxidation, *J. Power Sources*, **240**, 536–543.
106. Yang, G. T., Yang, X. K., Xu, M. L., Min, C. G., Xiao, H. F., Jiang, K. Z., Chen, L. Y., and Wang, G. H. (2013). Multi-walled carbon nanotube modified with methylene blue under ultraviolet irradiation as a platinum catalyst support for methanol oxidation, *J. Power Sources*, **222**, 340–343.
107. Hernández-Fernández, P., Baranton, S., Rojas, S., Ocón, P., Léger, J., and Fierro, J. L. G. (2011). Insights into the effects of functional groups on

carbon nanotubes for the electrooxidation of methanol, *Langmuir*, **27**, 9621–9629.

108. Hsieh, C., and Lin, J. (2009). Fabrication of bimetallic Pt–M (M = Fe, Co, and Ni) nanoparticle/carbon nanotube electrocatalysts for direct methanol fuel cells, *J. Power Sources*, **188**, 347–352.
109. Zhao, Y., Fan, L., Ren, J., and Hong, B. (2014). Electrodeposition of Pt–Ru and Pt–Ru–Ni nanoclusters on multi-walled carbon nanotubes for direct methanol fuel cell, *Int. J. Hydrogen Energy*, **39**, 4544–4557.
110. Huang, Y., Cai, J., Zheng, S., and Guo, Y. (2012). Fabrication of a high-performance Pb–PtCu/CNT catalyst for methanol electro-oxidation, *J. Power Sources*, **210**, 81–85.
111. Liao, S., Holmes, K., Tsaprailis, H., and Birss, V. I. (2006). High performance PtRuIr catalysts supported on carbon nanotubes for the anodic oxidation of methanol, *J. Am. Chem. Soc.*, **128**, 3504–3505.
112. Tokarz, W., Lota, G., Frackowiak, E., Czerwinski, A., and Piela, P. (2013). Fuel cell testing of Pt–Ru catalysts supported on differently prepared and pretreated carbon nanotubes, *Electrochim. Acta*, **98**, 94–103.
113. Kang, S., Lim, S., Peck, D., Kim, S., Jung, D., Hong, S., Jung, H., and Shul, Y. (2012). Stability and durability of PtRu catalysts supported on carbon nanofibers for direct methanol fuel cells, *Int. J. Hydrogen Energy*, **37**, 4685–4693.
114. Santasalo-Aarnio, A., Borghei, M., Anoshkin, I. V., Nasibulin, A. G., Kauppinen, E. I., Ruiz, V., and Kallio, T. (2012). Durability of different carbon nanomaterial supports with PtRu catalyst in a direct methanol fuel cell, *Int. J. Hydrogen Energy*, **37**, 3415–3424.
115. Zhang, C., Hu, J., Nagatsu, M., Shu, X., Toyoda, H., Fang, S., and Meng, Y. (2011). Magnetron sputtering of platinum nanoparticles onto vertically aligned carbon nanofibers for electrocatalytic oxidation of methanol, *Electrochim. Acta*, **56**, 6033–6040.
116. Khosravi, M., and Amini, M. K. (2010). Flame synthesis of carbon nanofibers on carbon paper: physicochemical characterization and application as catalyst support for methanol oxidation, *Carbon*, **48**, 3131–3138.
117. Sebastián, D., Suelves, I., Pastor, E., Moliner, R., and Lázaro, M. J. (2013). The effect of carbon nanofiber properties as support for PtRu nanoparticles on the electrooxidation of alcohols, *Appl. Catal. B*, **132–133**, 13–21.

118. Zainoodin, A. M., Kamarudin, S. K., Masdar, M. S., Daud, W. R. W., Mohamad, A. B., and Sahari, J. (2014). High power direct methanol fuel cell with a porous carbon nanofiber anode layer, *Appl. Energy*, **113**, 946–954.
119. Feng, C., Takeuchi, T., Abdelkareem, M. A., Tsujiguchi, T., and Nakagawa, N. (2013). Carbon–CeO2 composite nanofibers as a promising support for a PtRu anode catalyst in a direct methanol fuel cell, *J. Power Sources*, **242**, 57–64.
120. Wang, D., Liu, Y., Huang, J., and You, T. (2012). In situ synthesis of Pt/carbon nanofiber nanocomposites with enhanced electrocatalytic activity toward methanol oxidation, *J. Colloid Interf. Sci.*, **367**, 199–203.
121. Qin, Y., Yang, H., Zhang, X., Li, P., Zhou, X., Niu, L., and Yuan, W. (2010). Electrophoretic deposition of network-like carbon nanofibers as a palladium catalyst support for ethanol oxidation in alkaline media, *Carbon*, **48**, 3323–3329.
122. Chang, H., Joo, S. H., and Pak, C. (2007). Synthesis and characterization of mesoporous carbon for fuel cell applications, *J. Mater. Chem.*, **17**, 3078–3088.
123. Joo, S. H., Pak, C., You, D. J., Lee, S., Lee, H. I., Kim, J. M., Chang, H., and Seung, D. (2006). Ordered mesoporous carbons (OMC) as supports of electrocatalysts for direct methanol fuel cells (DMFC): effect of carbon precursors of OMC on DMFC performances. *Electrochim. Acta*, **52**, 1618–1626.
124. Salgado, J. R. C., Alcaide, F., Álvarez, G., Calvillo, L., Lázaro, M. J., and Pastor, E. (2010). Pt–Ru electrocatalysts supported on ordered mesoporous carbon for direct methanol fuel cell, *J. Power Sources*, **195**, 4022–4029.
125. Maiyalagan, T., Alaje, T. O., and Scott, K. (2012). Highly stable Pt–Ru nanoparticles supported on three-dimensional cubic ordered mesoporous carbon (Pt–Ru/CMK-8) as promising electrocatalysts for methanol oxidation, *J. Phys. Chem. C*, **116**, 2630–2638.
126. Bruno, M. M., Petruccelli, M. A., Viva, F. A., and Corti, H. R. (2013). Mesoporous carbon supported PtRu as anode catalyst for direct methanol fuel cell: polarization measurements and electrochemical impedance analysis of mass transport, *Int. J. Hydrogen Energy*, **38**, 4116–4123.
127. Alegre, C., Gálvez, M. E., Baquedano, E., Moliner, R., Pastor, E., and Lázaro, M. J. (2013). Oxygen-functionalized highly mesoporous carbon xerogel based catalysts for direct methanol fuel cell anodes, *J. Phys. Chem. C*, **117**, 13045–13058.

128. Viva, F. A., Bruno, M. M., Jobbagy, M., and Corti, H. R. (2012). Electrochemical characterization of PtRu nanoparticles supported on mesoporous carbon for methanol electrooxidation, *J. Phys. Chem. C*, **116**, 4097–4104.

129. Chen, D., Tang, L. H., and Li, J. H. (2010). Graphene-based materials in electrochemistry, *Chem. Soc. Rev.*, **39**, 3157–3180.

130. Huang, C. C., Li, C., and Shi, G. Q. (2012). Graphene based catalysts, *Energy Environ. Sci.*, **5**, 8848–8868.

131. Cao, Y. C., Xu, C., Wu, X., Wang, X., Xing, L., and Scott, K. (2011). A poly (ethylene oxide)/graphene oxide electrolyte membrane for low temperature polymer fuel cells, *J. Power Sources*, **196**, 8377–8382.

132. Liu, S., Wang, J., Zeng, J., Ou, J., Li, Z., Liu, X., and Yang S. (2010). "Green" electrochemical synthesis of Pt/graphene sheet nanocomposite film and its electrocatalytic property, *J. Power Sources*, **195**, 4628–4633.

133. Guo, S., Dong, S., and Wang, E. (2010). Three-dimensional Pt-on-Pd bimetallic nanodendrites supported on graphene nanosheet: facile synthesis and used as an advanced nanoelectrocatalyst for methanol oxidation, *ACS Nano*, **1**, 547–555.

134. Soin, N., Roy, S. S., Lim, T. H., and McLaughlin, J. A. D. (2011). Microstructural and electrochemical properties of vertically aligned few layered graphene (FLG) nanoflakes and their application in methanol oxidation, *Mater. Chem. Phys.*, **129**, 1051–1057.

135. Yao, Z. Q., Yue, R. R., Zhai, C. Y., Jiang, F. X., Wang, H. W., Du, Y. K., Wang, C. Y., and Yang, P. (2013). Electrochemical layer-by-layer fabrication of a novel three-dimensional Pt/graphene/carbon fiber electrode and its improved catalytic performance for methanol electrooxidation in alkaline medium, *Int. J. Hydrogen Energy*, **38**, 6368– 6376.

136. Zhao, J., Yu, H., Liu, Z., Ji, M., Zhang, L., and Sun, G. (2014). Supercritical deposition route of preparing Pt/graphene composites and their catalytic performance toward methanol electrooxidation, *J. Phys. Chem. C*, **118**, 1182–1190.

137. Zhong, J. P., Fan, Y. J., Wang, H., Wang, R. X., Fan, L. L., Shen, X. C., and Shi, Z. J. (2013). Highly active Pt nanoparticles on nickel phthalocyanine functionalized graphene nanosheets for methanol electrooxidation, *Electrochim. Acta*, **113**, 653–660.

138. Zhong, J. P., Fan, Y. J., Wang, H., Wang, R. X., Fan, L. L., Shen, X. C., and Shi, Z. J. (2013). Copper phthalocyanine functionalization of graphene nanosheets as support for platinum nanoparticles and their enhanced performance toward methanol oxidation, *J. Power Sources*, **242**, 208–215.

139. Li, Y., Gao, W., Ci, L., Wang, C., and Ajayan, P. M. (2010). Catalytic performance of Pt nanoparticles on reduced graphene oxide for methanol electro-oxidation, *Carbon*, **48**, 1124 -1130.

140. Sharma, S., Ganguly, A., Papakonstantinou, P., Miao, X., Li, M., Hutchison, J. L., Delichatsios, M., and Ukleja, S. (2010). Rapid microwave synthesis of CO tolerant reduced graphene oxide-supported platinum electrocatalysts for oxidation of methanol, *J. Phys. Chem. C*, **114**, 19459–19466.

141. Wang, S., Jiang, S. P., and Wang, X. (2011). Microwave-assisted one-pot synthesis of metal/metal oxide nanoparticles on graphene and their electrochemical applications, *Electrochim. Acta*, **56**, 3338–3344.

142. Qiu, J. D., Wang, G. C., Liang, R., Xia, X. H., and Yu, H. W. (2011). Controllable deposition of platinum nanoparticles on graphene as an electrocatalyst for direct methanol fuel cells, *J. Phys. Chem. C*, **115**, 15639–15645.

143. Ren, F. F., Wang, C. Q., Zhai, C. Y., Jiang, F. X., Yue, R. R., Du, Y. K., Yang, P., and Xu, J. K. (2013). One-pot synthesis of a RGO-supported ultrafine ternary PtAuRu catalyst with high electrocatalytic activity towards methanol oxidation in alkaline medium, *J. Mater. Chem. A*, **1**, 7255–7261.

144. Gao, L. N., Yue, W. B., Tao, S. S., and Fan, L. Z. (2013). Novel strategy for preparation of graphene-Pd, Pt composite, and its enhanced electrocatalytic activity for alcohol oxidation, *Langmuir*, **29**, 957–964.

145. Antolini, E., and Gonzalez, E. R. (2009). Polymer supports for low-temperature fuel cell catalysts, *Appl. Catal. A*, **365**, 1–19.

146. Patra, S., and Munichandraiah, N. (2009). Electrooxidation of methanol on Pt-modified conductive polymer PEDOT, *Langmuir*, **25**, 1732–1738.

147. Nagashree, K. L. Raviraj, N. H., and Ahmed, M. F. (2010). Carbon paste electrodes modified by Pt and Pt-Ni microparticles dispersed in polyindole film for electrocatalytic oxidation of methanol, *Electrochim. Acta*, **55**, 2629–2635

148. Stanis, R. J., Lambert, T. N., and Yaklin, M. A. (2010). Poly(3,4-ethylenedioxythiphene) (PEDOT)-modified anodes: reduced methanol crossover in direct methanol fuel cells, *Energy Fuels*, **24**, 3125–3129.

149. Habibi, B., and Pournaghi-Azar, M. H. (2010). Methanol oxidation on the polymer coated and polymer-stabilized Pt nano-particles: a comparative study of permeability and catalyst particle distribution

ability of the PANI and its derivatives, *Int. J. Hydrogen Energy*, **35**, 9318–9328.

150. Huang,Y. F., Lin, C. W., Chang, C. S., and Ho, M. J. (2011). Alternative platinum electrocatalyst supporter with micro/nanostructured polyaniline for direct methanol fuel cell applications, *Electrochim. Acta*, **56**, 5679–5685.
151. Yuan, X. X., Ding, X. L., Wang, C. Y., and Ma, Z. F. (2013). Use of polypyrrole in catalysts for low temperature fuel cells, *Energy Environ. Sci.*, **6**, 1105–1124.
152. Selvaraj, V., and Alagar, M. (2008) Ethylene glycol oxidation on Pt and Pt–Ru nanoparticle decorated polythiophene/multiwalled carbon nanotube composites for fuel cell applications, *Nanotechnology*, **19**, 045504/1–8.
153. Habibi, E., and Razmi, H. (2013). Kinetics of direct ethanol fuel cell based on Pt–PoPD nano particle anode catalyst, *Int. J. Hydrogen Energy*, **38**, 5442–5448.
154. Wang, L., Wang, Y., Li, A., Yang, Y., Tang, Q., Cao, H., Qi, T., and Li, C. (2014). Electrocatalysis of carbon black– or poly(diallyldimethylammonium chloride)–functionalized activated carbon nanotubes–supported Pd–Tb towards methanol oxidation in alkaline media, *J. Power Sources*, **257**, 138–146.
155. Gao, H., He, J. B., Wang, Y., and Deng, N. (2012). Advantageous combination of solid carbon paste and a conducting polymer film as a support of platinum electrocatalyst for methanol fuel cell, *J. Power Sources*, **205**, 164–172.
156. Arbizzani, C., Biso, M., Manferrari, E., and Mastragostino, M. (2008). Methanol oxidation by pEDOT–pSS/PtRu in DMFC, *J. Power Sources*, **178**, 584–590.
157. Tintula, K. K., Pitchumani, S., Sridhar, P., and Shukla, A. (2010). A solid–polymer electrolyte direct methanol fuel cell (DMFC) with Pt–Ru nanoparticles supported onto poly(3,4–ethylenedioxythiophene) and polystyrene sulphonic acid polymer composite as anode, *J. Chem. Sci.*, **122**, 381–389.
158. Hoa, L. Q., Yoshikawa, H., Saito, M., and Tamiya, E. (2012). Study of co–assembled conducting polymers for enhanced ethanol electrooxidation reaction, MRS Online Proceedings Library, 1446.
159. Jia, Y., Jiang, J., Sun, K., and Dai, T. (2012). Electrocatalytic performance of Pt supported on polyaniline–poly(styrene sulfonate) hydrogel, *J. Appl. Polym. Sci.*, **125**, 3702–3707.

160. Antolini, E. (2010). Composite materials: an emerging class of fuel cell catalyst supports, *Appl. Catal. B*, **100**, 413–426.

161. Guo, D. J., and Jing, Z. H. (2011). Electrocatalytic properties of platinum nanoparticles supported on fluorine tin dioxide/multi-walled carbon nanotube composites for methanol electrooxidation in acidic medium, *J. Colloid Interface Sci.*, **359**, 257–260.

162. Guo, D. J., Qiu, X. P., Zhu, W. T., and Chen, L. Q. (2009). Synthesis of sulfated ZrO2/MWCNT composites as new supports of Pt catalysts for direct methanol fuel cell application, *Appl. Catal. B*, **89**, 597–601.

163. Zhao, Z., Fang, X., Li, Y., Wang, Y., Shen, P. K., Xie, F., and Zhang, X. (2009). The origin of the high performance of tungsten carbides/carbon nanotubes supported Pt catalysts for methanol electrooxidation, *Electrochem. Commun.*, **11**, 290–293.

164. Wang, K., Wang, Y., Liang, Z., Liang, Y., Wu, D., Song, S., and Tsiakaras, P. (2014). Ordered mesoporous tungsten carbide/carbon composites promoted Pt catalyst with high activity and stability for methanol electrooxidation, *Appl. Catal. B*, **147**, 518–525.

165. Karakoti, A. S., King, J. E. S., Vincent, A., and Seal, S. (2010). Synthesis dependent core level binding energy shift in the oxidation state of platinum coated on ceria–titania and its effect on catalytic decomposition of methanol, *Appl. Catal. A*, **388**, 262–271.

166. Mikhaylova, A. A., Tusseeva, E. K., Mayorova, N. A., Rychagov, A. Yu., Volfkovich, Yu. M., Krestinin, A. V., and Khazova, O. A. (2011). Single-walled carbon nanotubes and their composites with polyaniline. Structure, catalytic and capacitive properties as applied to fuel cells and supercapacitors, *Electrochim. Acta*, **56**, 3656–3665.

167. Wei, L., Fan, Y. J., Ma, J. H., Tao, L. H., Wang, R. X., Zhong, J. P., and Wang, H. (2013). Highly dispersed Pt nanoparticles supported on manganese oxide–poly(3,4-ethylenedioxythiophene)–carbon nanotubes composite for enhanced methanol electrooxidation, *J. Power Sources*, **238**, 157–164.

168. Yue, R., Zhang, Q., Wang, C., Du, Y., Yang, P., and Xu, J. (2013). Graphene–poly(5-aminoindole) composite film as Pt catalyst support for methanol electrooxidation in alkaline medium, *Electrochim. Acta*, **107**, 292–300.

169. Xia, Y., Xiong, Y. J., Lim, B., and Skrabalak, S. E. (2009). Shape-controlled synthesis of metal nanocrystals: simple chemistry meets complex physics, *Angew. Chem., Int. Ed.*, **48**, 60–103.

170. Guo, S. J., and Wang, E. K. (2011). Noble metal nanomaterials: controllable synthesis and application in fuel cells and analytical sensors, *Nano Today*, **6**, 6240–6264.

171. Zhou, Z. Y., Tian, N., Li, J. T., Broadwell, I., and Sun, S. G. (2011). Nanomaterials of high surface energy with exceptional properties in catalysis and energy storage, *Chem. Soc. Rev.*, **40**, 4167–4185.

172. Zhou, Z. Y., Tian, N., Huang, Z. Z., Chen, D. J., and Sun, S. G. (2008). Nanoparticle catalysts with high energy surfaces and enhanced activity synthesized by electrochemical method, *Faraday Discuss.*, **140**, 81–92.

173. Lim, S. I., Ojea-Jiménez, I., Varon, M., Casals, E., Arbiol, J., and Puntes, V. (2010). Synthesis of platinum cubes, polypods, cuboctahedrons, and raspberries assisted by cobalt nanocrystals, *Nano Lett.*, **10**, 964–973.

174. Kang, Y. J., and Murray, C. B. (2010). Synthesis and electrocatalytic properties of cubic Mn–Pt nanocrystals (nanocubes), *J. Am. Chem. Soc.*, **132**, 7568–7569.

175. Gong, X. Z., Yang, Y., and Huang, S. M. (2011). Mn3O4 catalyzed growth of polycrystalline Pt nanoparticles and single crystalline Pt nanorods with high index facets, *Chem. Commun.*, **47**, 1009–1011.

176. Tian, N., Zhou, Z. Y., and Sun, S. G. (2009). Electrochemical preparation of Pd nanorods with high-index facets, *Chem. Commun.*, **12**, 1502–1504.

177. Tsai, H. Y., Shih, Z. Y., Lin, Z. H., and Chang, H. T. (2013). Te/Pt nanonetwork modified carbon fiber microelectrodes for methanol oxidation, *Nanotechnology*, **24**, 195402–195409.

178. Song, Y. J., Han, S. B., and Park, K. W. (2010). Pt nanowire electrodes electrodeposited in PVP for methanol electrooxidation, *Mater. Lett.*, **64**, 1981–1984.

179. Zhu, C. Z., Guo, S. J., and Dong, S. J. (2012). Facile synthesis of trimetallic AuPtPd alloy nanowires and their catalysis for ethanol electrooxidation, *J. Mater. Chem.*, **22**, 14851–14855.

180. Mohanty, A., Garg, N., and Jin, R. C. (2010). A universal approach to the synthesis of noble metal nanodendrites and their catalytic properties, *Angew. Chem., Int. Ed.*, **49**, 4962–4966.

181. Zhang, H. M., Zhou, W. Q., Du, Y. K., Yang, P., and Wang, C. Y. (2010). One-step electrodeposition of platinum nanoflowers and their high efficient catalytic activity for methanol electro-oxidation, *Electrochem. Commun.*, **12**, 882–885.

182. Wei, L., Fan, Y. J., Wang, H. H., Tian, N., Zhou, Z. Y., and Sun, S. G. (2012). Electrochemically shape-controlled synthesis in deep eutectic solvents of Pt nanoflowers with enhanced activity for ethanol oxidation, *Electrochim. Acta*, **76**, 468–474.
183. Tian, N., Zhou, Z. Y., Sun, S. G., Ding, Y., and Wang, Z. L. (2007). Synthesis of tetrahexahedral platinum nanocrystals with high-index facets and high electro-oxidation activity, *Science*, **316**, 732–735.
184. Sau, T. K., and Rogach, A. L. (2010). Nonspherical noble metal nanoparticles: colloid-chemical synthesis and morphology control, *Adv. Mater.*, **22**, 1781–1804.
185. Huang, X. Q., Zhao, Z. P., Fan, J. M., Tan, Y. M., and Zheng, N. F. (2011). Amine-assisted synthesis of concave polyhedral platinum nanocrystals having {411} high-index facets, *J. Am. Chem. Soc.*, **133**, 4718–4721.
186. Xia, B. Y., Wu, H. B., Wang, X., and Lou, X. W. (2013). Highly concave platinum nanoframes with high-index facets and enhanced electrocatalytic properties, *Angew. Chem. Int. Ed.*, **52**, 12337–12340.
187. Wei, L., Fan, Y. J., Tian, N., Zhou, Z. Y., Zhao, X. Q., Mao, B. W., and Sun, S. G. (2012). Electrochemically shape-controlled synthesis in deep eutectic solvents—a new route to prepare Pt nanocrystals enclosed by high-index facets with high catalytic activity, *J. Phys. Chem. C*, **116**, 2040–2044.
188. Liang, Z. X., Shi, J. Y., Liao, S. J., and Zeng, J. H. (2010). Noble metal nanowires incorporated Nafion membranes for reduction of methanol crossover in direct methanol fuel cells, *Int. J. Hydrogen Energy*, **35**, 9182–9185.
189. Sakthivel, M., Schlange, A., Kunz, U., and Turek, T. (2010). Microwave assisted synthesis of surfactant stabilized platinum/carbon nanotube electrocatalysts for direct methanol fuel cell applications, *J. Power Sources*, **195**, 7083–7089.
190. Madaeni, S. S., Amirinejad, S., and Amirinejad, M. (2011). Phosphotungstic acid doped poly(vinyl alcohol)/poly(ether sulfone) blend composite membranes for direct methanol fuel cells, *J. Membrane Sci.*, **380**, 132–137.
191. Molla, S., and Compan, V. (2011). Performance of composite Nafion/PVA membranes for direct methanol fuel cells, *J. Power Sources*, **196**, 2699–2708.
192. Kulikovsky, A. A. (2012). A model for optimal catalyst layer in a fuel cell, *Electrochim. Acta*, **79**, 31–36.

193. Antolini, E. (2011). An empirical model to evaluate the contribution of alloyed and non-alloyed tin to the ethanol oxidation reaction on Pt-Sn/C catalysts based on the presence of SnO2 and a Pt(1-x)Snx solid solution: application to DEFC performance, *Int. J. Hydrogen Energy*, **36**, 11043–11047.

194. Avgouropoulos, G., Ioannides, T., Kallitsis, J. K., and Neophytides, S. (2011). Development of an internal reforming alcohol fuel cell: concept, challenges and opportunities, *Chem. Eng. J.*, **176–177**, 95–101.

195. Wu, Q. X., Zhao, T. S., Chen, R., and An, L. (2013). A sandwich structured membrane for direct methanol fuel cells operating with neat methanol, *Appl. Energy*, **106**, 301–306.

Chapter 2

Multimetallic Nanocatalysts for Anodic Reaction in Direct Alcohol Fuel Cell

Jayati Datta
Electrochemistry and Fuel Cell Laboratory, Department of Chemistry, Indian Institute of Engineering Science and Technology, Shibpur, Howrah 711 103, West Bengal, India
jayati_datta@rediffmail.com

The hydrogen fueled proton exchange membrane fuel cells (PEMFCs) have unique characteristics of high energy and zero emission. However, in the 21st century it has been a growing perception that electricity generation by H_2–O_2 fuel cell primarily depends on the concept of "Hydrogen Economy." On the other hand, aliphatic alcohols like methanol, ethanol, propanol, and butanol are characterized by their octane rating and are identified as the future choices for reducing oil imports. Among these, there has been a great deal of interest in the development of direct ethanol fuel cell (DEFC) technology operating on bioethanol as the renewable source that can easily be produced, on a large scale, from agricultural and municipal wastes. In fact, use of ethanol reduces carbon dioxide emission by an average of 34%, compared to gasoline. The main prerequisite of the liquid fuel for PEMFC application is an appreciably high electrochemical reactivity at

Nanomaterials for Direct Alcohol Fuel Cell
Edited by Yixuan Wang

ISBN 978-981-4669-00-9 (Hardcover), 978-981-4669-01-6 (eBook)
www.panstanford.com

relatively low temperature, which can be routed through judicious choice of stringent catalysts for alcohol oxidation as well as the oxygen reduction reaction. In the present chapter, a comprehensive discussion has been made on the state-of-the-art catalyst technology for anodic oxidation of alcohol and ethanol, in particular, for low temperature fuel cell application. In DEFC, the catalysts designed with multimetallic framework have shown improved functional properties resulting from the combinatorial effect of M–M inter-atomic distance, number of nearest neighboring metal atoms, d band vacancy, size, dispersion, and metal content on the surface. A great deal of interest has arisen in the DEFCs' working in alkaline environment, which allows the use of inexpensive non-platinum metal as well as the transitional metals and metal oxides as critical electrode components harnessing more energy and at the same time ensuring affordability of the fuel cell system.

2.1 Introduction

The 21st century has been witnessing serious threat to the progress of human civilization due to dearth of energy. The increasing energy demands and diminishing energy security have ultimately evolved several energy strictures and economic imperatives that are to be addressed by the entire world, so as to maintain the eco-sustainability for the future generations. These uncertainty problems can be addressed through the development of new materials, devices, and systems that aim to establish and utilize clean and renewable energies across the board spectrum of technological fields. In recent times, nanotechnology is being heralded as the most promising technology that would enable the remodeling of the past technological practices and lead to eco-friendly products for present-day end users. In the context of eco-sustenance of the future society, judicious adoption of controlled technologies always implies ethical application of materials and intelligent exploitation of available resources. In this respect the fuel cell technology has been shown to be very an interesting and promising alteration to solve the problem of clean electric power generation with high efficiency.

2.2 Low-Temperature Fuel Cells

Fuel cells are designed in different types and sizes so as to achieve the best possible electrochemical efficiency and the components are tailor made according to their compatibility with the existing energy utilization pattern. In the category of low-temperature fuel cells using hydrogen, the polymer electrolyte membrane fuel cell (PEMFC) is commonly claimed to be the most befitting device for portable and residential application [1, 2]. In most cases, the hydrogen used for PEMFCs operation is from the external reforming process, leading to a bulkier and heavier system and increasing simultaneously its complexity and cost [3]. Moreover, although low-temperature operation is one of the advantages of PEMFC, the fuel requirement is very stringent. Once the CO content in the hydrogen-rich gas is more than 10 ppm, an obvious decay in the cell performance will be observed, and to overcome this, high Pt catalyst loadings are usually required. H_2 also poses dangers that lead to critical safety issues related to its generation, handling, and storage. Options for on-board hydrogen storage include compressed gas, cryogenic (liquid) hydrogen, hydride storage, and carbon nanotubes [4]. However, there are drawbacks to each of these options and the storage problems with H_2 have spawned great interest in the identification of possible alternative fuels.

2.3 Fuel Flexibility

In order to circumvent the yet unsolved problems associated with the production, storage, and transfer of hydrogen fuel, the direct oxidation fuel cell (DOFC) has now emerged as a promising technology in the fuel cell arena. In contrast to the use of hydrogen, DOFCs have the liberty to use organic and inorganic fuels such as alcohols, carbohydrates, hydrazine, borohydrides, etc., which is directly fed in vapor or liquid phase into the anode compartment while air or oxygen is fed into the cathode compartment. Such fuel cells commonly use polymer electrolyte membrane (PEM) in acid environment, whereas in alkali environment the fuel cell can operate with or without the use of membrane. Use of liquid fuels can provide considerable weight and volume advantages over the external reforming PEMFC. Because of fuel flexibility, fuel cells can promote energy diversity and a transition to renewable energy sources.

Since aggressive research has been undertaken for the development of DOFC and direct alcohol fuel cell (DAFC) in particular, the latter can be classified according to the electrolyte and membrane used in the cell.

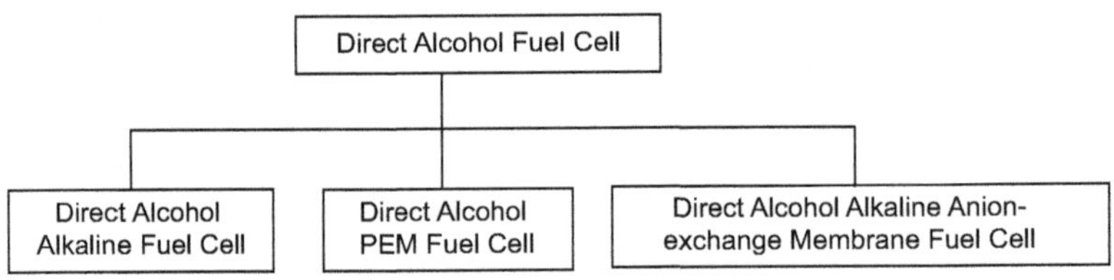

DAFC is now in the forefront, receiving increased attention due to its much simpler peripheral units than those of the H_2–O_2 PEMFC systems such as easy fuel delivery and storage, no need for reforming, high energy conversion efficiency, humidification and stack cooling, and favorable power capability compared with other fuel cell systems [5]. Among the three categories, DAFCs based on proton exchange membranes is the most studied one. DAFCs usually have compact design and potentially can offer up to 10 times the energy density of rechargeable batteries [6]. In addition, DAFCs can be designed to operate at low temperature, which significantly reduces thermal management challenges for small systems. These advantages make the technology attractive to the rapidly growing need for portable power sources and in transport sectors.

However, the low performance of DA (PEM) FC is due to the sluggish alcohol oxidation kinetics in acidic medium, alcohol crossover through PEM, and, consequently, cathode polarization. It has been reported that improved alcohol oxidation kinetics can be facilitated in an alkali analogue of DA (PEM) FC [7, 8]. However, the only problem in using an alkaline electrolyte is carbonate formation. Thus, to avoid the carbonation, use of alkaline anion exchange membrane (AAEM) is advocated and such systems have shown reasonable performance stability [9].

2.4 Selection of Fuels for Direct Alcohol Fuel Cell

The alcohols can be used as the hydrogen carriers in fuel cells. The most commonly used alcohols in DAFCs are methanol and ethanol. Although the electrochemical reactivity of ethanol is comparable to that of the methanol on platinum-based catalyst, the complete electro-oxidation of C_2 molecule to CO_2 involving C–C bond rupture remains in question. Similar to methanol, ethanol also produces CO as a reaction intermediate which ultimately poisons the catalyst surface. Among the other possible alcohol candidates, 2-propanol, which has high energy density, has also been investigated as a fuel for DAFCs [10]. Due to the characteristic molecular structure, 2-propanol is the only molecule among the C_3 alcohol that does not dissociate into CO at the surface of the electrocatalyst. The DAFCs using 2-propanol fuel show moderate performance and a much lower cross-over current compared to those obtained with DMFC [7, 11]. However, the position of the OH group at the C_α–position possibly makes the molecule much more stable, so that the C–C bond scission become unfavorable, leading only to dehydrogenation [10].

2.5 Ethanol: A Green Fuel

Among all the possible liquid fuels investigated up to a few years ago, methanol had been the choice due to its relatively higher electrochemical activity compared to other liquid fuels. However, it is well known that methanol is volatile and relatively toxic. Instead, ethanol has emerged as a green fuel since it is eco-friendly and has more energy density compared to methanol (8.01 kWh kg^{-1} vs. 6.09 kWh kg^{-1}) [12].

Moreover, bioethanol is a renewable energy source and can be easily produced in great quantity by the fermentation of sugar containing raw materials from agriculture and even from the organic fraction of municipal solid wastes. Furthermore, use of bioethanol in fuel cell has the significant advantage of being nearly CO_2 neutral since the carbon dioxide produced in the process is consumed for biomass growth, offering a nearly closed carbon loop (Fig. 2.1).

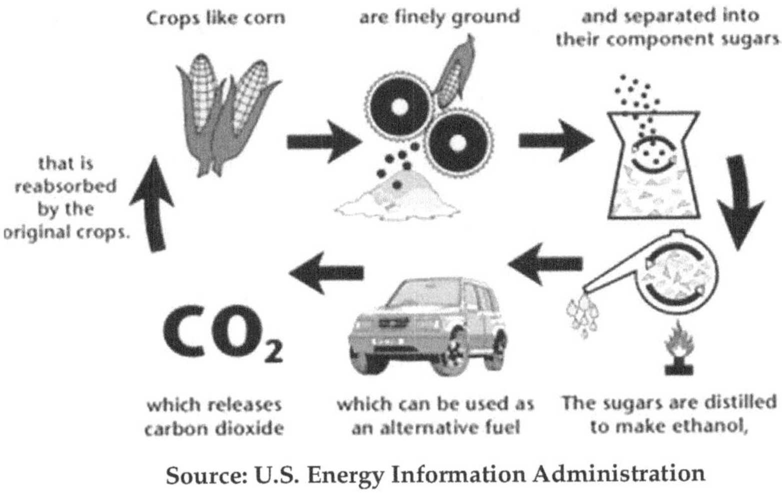

Figure 2.1 Carbon dioxide recycles with ethanol fuel.

Ethanol can therefore open up the possibility of replacing methanol in a DAFC.

However, one of the chief hiatus in the commercialization of DEFC is still its low power density compared to the hydrogen fuel cell. The major problems associated with the CO poisoning of the anode catalyst in DEFC have been addressed through successful attempts in developing multimetallic nanocatalysts in their alloyed or composite form involving the precious metals as well as the transitional metal oxides which bring about the synergy due to the intercalating nature of the oxide components into the metallic matrix [13]. In order to bring down the cost of fuel cell stack component, a greater challenge ahead is the search for non-platinum catalyst formulations, which are already on satisfactory trials [14].

2.6 Selection of Electrocatalyst Materials for Oxidation of Alcohols

The nature and the structure of the electrode materials play a key role in the adsorption and electro-oxidation of aliphatic alcohols.

The use of proton exchange membrane as acid electrolyte induces pH limitations, and therefore, in such a strong acid environment, the only stable and active catalyst for the electro-oxidation of alcohols is perhaps platinum. The dissociative chemisorption [15] of alcohols like methanol and ethanol leads to the formation of strongly poisoning species (adsorbed CO), which block the electro-active sites of the catalyst, resulting in the high overpotentials observed in the electro-oxidation of these alcohols. Another catalytic challenge is the breaking of C–C bond in the higher homologous series of alcohols. None of the individual precious metal is capable of replacing platinum for the oxidation reaction because of the relative competence of adsorption of organic molecules as well as kinetic and corrosion restrictions, particularly in acidic medium. Although Pt shows outstanding electrocatalytic behavior towards a large number of charge transfer phenomena, its high cost and limited supply is a major concern. With the significant rise in the price of platinum group of metals, the cost of membrane electrode assembly, MEA (comprising catalyst layers and the membrane) and the fuel cell as a whole has sharply increased (Fig. 2.2). The application of alkali electrolytes to DAFCs has been effective for the reduction of Pt in catalyst loading, and it allows the use of less expensive non-precious metal in catalyst matrices [16]. In this circumstance, attempts for

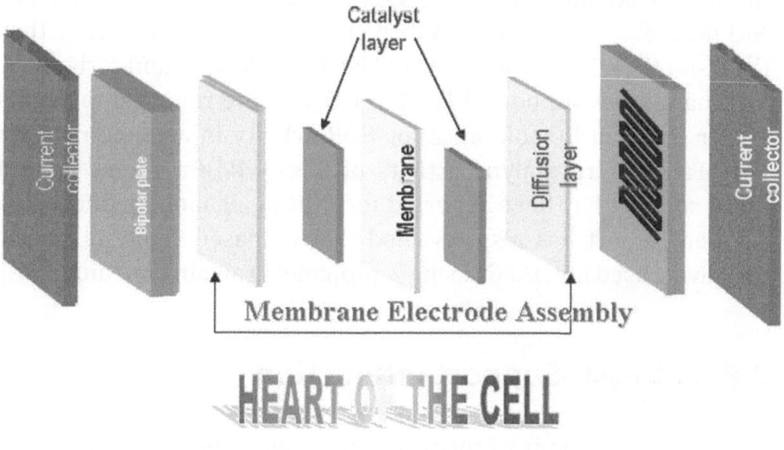

Figure 2.2 Exploded view of the components of a single cell.

replacement or the reduction in consumption of platinum catalysts is an urgent issue that clearly demands exploring the use of other metals, e.g., Ru, Pd, Rh, Au, Ni, Co, Sn, Mo, W, and so on, at various levels, which could lower the consumption of platinum considerably without compromising the efficiency [17, 18]. In this context, it may be noted that the estimated recoverable resources of platinum are about 40,000 ton, and those of palladium, for example, are 50 times higher [19].

2.7 Binary Systems Studied for DAFC

Among the different binary nanoparticles (NPs) investigated in acidic medium, PtRu and PtSn appeared to be less poisoned by adsorbed intermediates (CO) generated from the dissociation of methanol and ethanol [20, 21]. However, the PtRu anode catalyst does not show much electro-activity towards oxidation of higher alcohols like ethanol and 2-propanol, which suffer from difficulties associated with complete oxidation of alcohols to CO_2 [10]. A potential anode material for alkaline DAFC was developed by Sengupta et al. with PtRu alloyed NPs on Cu:Ni alloy support [22]. Improved electrocatalytic activity was obtained by promoting Pd/C with nanocrystalline oxides like CeO_2, Co_3O_4, Mn_3O_4, and NiO in the use of ethanol oxidation reaction (EOR) [23]. The CeO_2 and NiO promoted Pd/C catalysts displayed superior EOR activity than the Pt-Ru/C (E-TEK). NiO-activated Pd/C (Pd-NiO/C) and carbonized TiO_2 nanotubes supported Pd (Pd/TiO_2C) were reported to exhibit higher levels of CO tolerance for EOR activity in alkaline medium [24]. Fair electrocatalytic activity of binary Pd-Ru anode catalyst on Ni support has been reported for EOR in alkaline medium [25]. A recent report has also revealed that Pd-based catalysts can be effectively used for DAFCs using 2-propanol in alkaline medium [26].

2.8 Catalyst–Support Interaction

In recent times, many of the promising approaches for reduction in Pt loading in fuel cells have gained considerable importance through supporting tactics [27]. The metal support interactions have been considered to be effective in the growth, the structure formulation, and the dispersion of metallites or alloyed NPs on carbon supports, thus providing potential benefits to the improvement of the catalytic properties and stability of the electrocatalyst. In addition to the development of a variety of nanocatalysts, the use of carbon support turns out to be critical for providing fine dispersion, better utilization, increased activity, and stability of the nanoparticles. The support materials have strong influence on the properties of noble metal catalysts, such as metal particle size, morphology, size distribution, degree of alloying, stability, and dispersion [28]. At the same time, in fuel cells, they can also affect the electronic conductivity, electrochemical active area, and the mass transport phenomenon during the operation. In general, carbon black in the form of Vulcan XC-72 with controlled micropores or mesopores, inherits most of the above properties as discussed and remains to be the choice as good catalyst support, with comparatively less price among the others, for low-temperature fuel cells.

However, a few disadvantages are also associated with the use of carbon black as support material. In this respect, several other supporting materials are being deployed for use in fuel cells, such as CNF, CNT, TiO_2, carbon microsphere, diamond, fullerenes, and carbon fibers [29]. Organic conducting polymers having synergetic or complementary behaviors between polymer and inorganic matrices are being explored as substrate materials for catalytic enhancement in low-temperature fuel cells [30].

2.9 Synthesis of Nanoparticles

2.9.1 Chemical Route

Preparative strategies play an important role in obtaining tailor-made NPs as catalysts for fuel cell reactions. The common criteria for a high performance catalyst are (i) a narrow nanoscale size distribution, (ii) a uniform composition throughout the nanoparticles, (iii) a fully alloyed degree, and (iv) high dispersion on the support materials. According to these criteria, some of the cost-effective methods have shown promise for reaching performance optimization by controlling synthetic procedures and conditions. There are two major routes, through which catalyst are synthesized: chemical and electrochemical routes. The chemical methods include mainly the impregnation, colloidal, and microemulsion techniques [31–33]. Out of these, the impregnation method is a simple and straightforward preparation technique widely used for the Pt-based catalyst preparation. During the impregnation step, metal precursors are mixed with high-surface-area carbon black in aqueous solution to form a homogeneous mixture. The chemical reduction step can be carried out by liquid phase reduction, using any one of the reducing agents like $Na_2S_2O_3$, $NaBH_4$, $Na_4S_2O_5$, N_2H_4, or formic acid. During the impregnation process, many factors can affect the composition, morphology, and dispersion of a catalyst, resulting in the variation of catalytic activity.

However, with the above conventional methods, sometimes it is difficult to achieve tailor-made sizes of Pt nanoparticles within 2–5 nm range with homogeneous dispersion on the carbon support. Particularly, reducing the particle size down to 5 nm is still a big challenge for these methods. Hence, in order to tailor the size of the nanoparticles with uniform dispersion on the support materials, the synthetic procedure needs modifications, e.g., by the use of stabilizing agents like surfactants, ligands, polymers, and others [34, 35]. The use of capping agents like tetraalkylammonium salts (NR_4^+), triphenylphosphine (PPh_3), polyvinyl pyrrolidine (PVP), and polyvinyl alcohol (PVA) has drawn attention, because during synthesis narrow-size distribution can be achieved through stabilization (Fig. 2.3) either by steric hindrance, provided by shielding the metal core with organic chain molecules, or by electrostatic charges provided by oppositely charged tetra alkyl ammonium ions (NR_4^+).

Singh et al. [36] have shown that almost half of the loading of precious Pt is required in the modified catalyst to maintain the same level of reactivity when compared with the nonprotected catalyst. This becomes one of the desirable features associated with the use of stabilizing agent which offers nano-size distribution as well as controlled dispersion of metal NPs throughout the catalyst matrix. Essentially, these reagents are capable of projecting the active planes

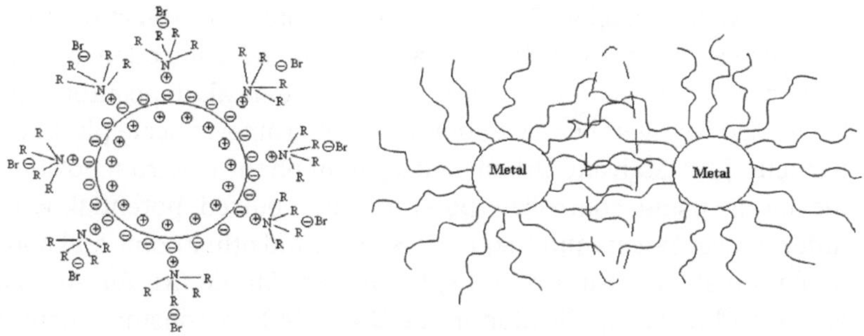

Figure 2.3 (a) Electrostatic charges. (b) Steric hindrance.

of the nanocrystallites as favorable sites for electro-oxidation of alcohols. This in fact complies with the challenge of high utilization of precious metals at minimum possible loading during the fuel cell reaction. The coverage of the metal particles by large lypophillic alkyl groups exerts a remarkable stability to the catalyst particles. Bimetallic PtRu nanocatalysts were synthesized by a novel method using PVP as the stabilizing agent. Progressive incorporation of Ru induces structural changes and control the size distribution of the NPs [37].

2.9.2 Electrosynthesis

Electrodeposition involves application of constant current/potential or linear variation of the same to an electrode immersed in a precursor solution so as to attain the threshold potential at which reduction can occur. The surface morphology of the substrate material is one of the key parameters to obtain quality deposition. It is more economic and facile than the other types of deposition techniques like chemical vapor deposition, cladding, sputtering, and spin coating [38].

Electrodeposition occurs in two distinct stages, nucleation and growth on the preferential sites of the substrate. During nucleation the atoms formed from precursor solution regroup themselves to form cores or nuclei, which are followed by the growth process during which the atoms from the solution are directly incorporated in the existing nuclei. For the reduction step, the reaction time and hence the growth of the particle can be controlled by the amount of the current passed. Electrochemical deposition has been used in several cases to obtain submicrometer-scale platinum particles [39]. Stab and Urban filed a patent on the electrodeposition technique using controlled potential and pulsed potential with different waveforms and frequencies for Pt and other noble and non-noble metals like Ru, Rh, Ir, Re, Os, Pd, Pb, Sn, Co, Ni, Au, Ag, Cu, W, Zn, and Cr for application in the DMFC [40]. A constant current deposition technique was adopted by Datta and Sengupta [41] for synthesizing platinum NPs on graphite support. These platinum catalysts were subjected to electrocatalysis towards EOR in order to assess (i) the influence of the deposition current density on the surface morphology of platinized carbon electrodes, (ii) the relationship between platinum loading and roughness factor of the deposited Pt catalysts, and (iii) the effect of platinum dispersion on the electrocatalytic activity towards ethanol electro-oxidation. A clear trade-off between the amount of platinum loading and the roughness factor of the electrodes has been shown in their report using an optimum current density of 3 mA cm^{-2}. They further demonstrated that controlled growth and narrow size distribution (Fig. 2.4) can be attained with polytetrafluroethylene (PTFE) polymer used in the deposition bath resulting in highly porous Pt matrix.

(a) (b)

Figure 2.4 SEM images of electrodeposited (a) platinum on graphite and (b) PTFE-modified Pt/C.

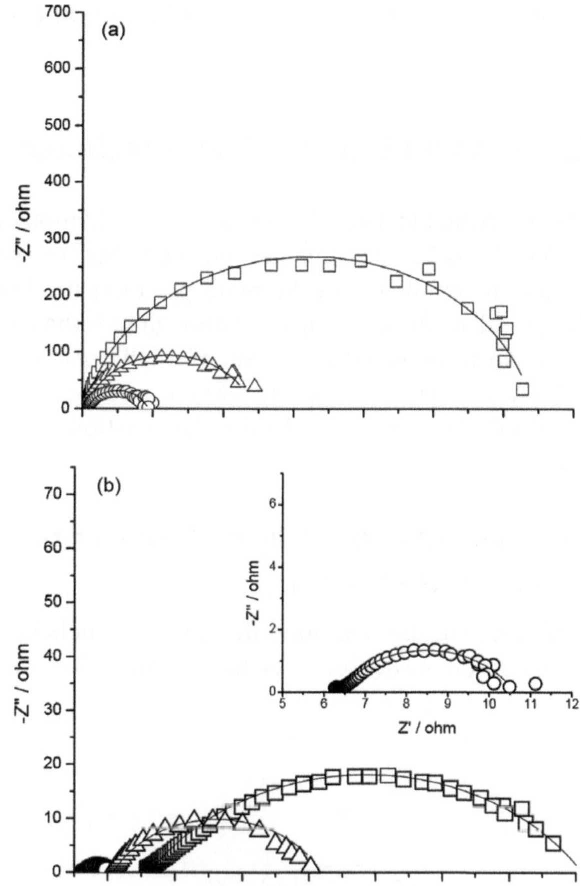

Figure 2.5 Impedance spectra of (a) C/Pt and (b) C/Pt(PTFE) electrodes in 1.0 M ethanol and 0.5 M H_2SO_4 at 0.7 V. Deposition current densities: 3 mA cm^{-2} (○), 5 mA cm^{-2} (Δ), and 10 mA cm^{-2} (□).

Further, by employing electrochemical impedance spectroscopy a significant lowering of the charge transfer resistance in EOR was observed for the PTFE-modified platinum catalysts, which indicated enhancement in the reaction kinetics (Fig. 2.5). These deposition conditions, therefore, allow optimization of the platinum NPs forming the matrix accessible to the reacting molecules because of the microroughness of the platinum particles. Sengupta et al. and Mahapatra et al., in their further work, electrodeposited a series of

Pt-based binary and ternary alloys for application in EOR studies [42, 43].

2.10 Catalysts for DEFC: A Green Technology

The desired reaction in a DEFC is the complete oxidation of ethanol to CO_2, which yields 12 electrons per molecule, but the reaction on Pt is slow and incomplete and the main products are likely to be acetaldehyde and acetic acid [10, 21]. Although Pt is the best-known catalyst for adsorption and dissociation of small organic molecules, it has the limitation of breaking the C–C bond and is readily poisoned by adsorbed CO. The anodic and cathodic reactions in DEFC are shown in Fig. 2.6.

2.10.1 Nanomaterials Specially Focused on EOR Studies: Influence of Adatoms

The rate of breaking the C–C bond in higher alcohols like ethanol strongly depends on the degree of its interaction with the adsorption

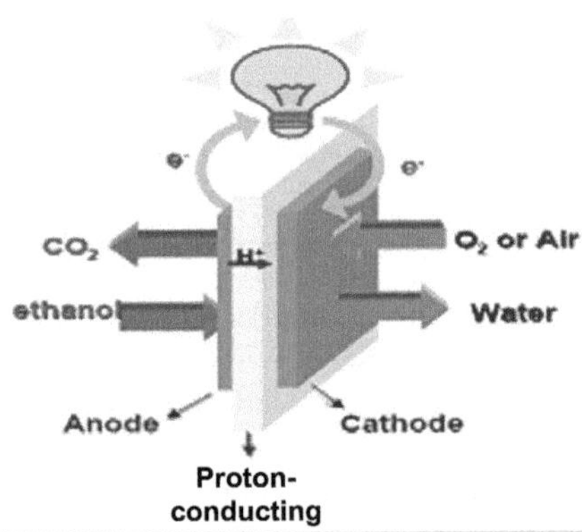

Anode: $C_2H_5OH + 3H_2O \rightarrow 2CO_2 + 12H^+ + 12e$ $E^0 = 0.085$ V

Cathode: $3O_2 + 12H^+ + 12e \rightarrow 6H_2O$ $E^0 = 1.23$ V

Overall: $C_2H_5OH + 3O_2 \rightarrow 2CO_2 + 3H_2O$ $E = 1.145$ V

Figure 2.6 Schematic of DEFC.

sites of the catalyst and the kinetics of electron transfer. Hence to overcome these problems, formulation of stringent electrocatalyst in low-temperature fuel cells has become the subject of extensive research. In fact, the design of multimetallic alloyed catalysts stems from exploring the synergistic properties as a result of the change in the surface binding sites, the lattice changes, and the electronic effects for enhancing the catalytic activity, improving the stability and ultimately reducing the use of Pt in the catalysts. The catalysts in their alloyed state having lower d-band electron density than Pt are capable of weakening the bond between the strongly adsorbed intermediate and Pt by the negative inductive effect, often known as the "ligand effect" [44]. Further studies have involved a second or multiple choice of metals which have greater ability to form surface oxide species compared to Pt, commonly guided by the "bifunctional mechanism" [45]. These combined or alloyed metal crystallites are being advocated as "CO-tolerant" catalysts for low-temperature direct oxidation fuel cells (DOFCs) in general.

Among the primary characterization techniques for nanomaterials, X-ray diffraction is often used to obtain a quick estimate of the grain size, lattice parameters, and the extend of alloying for multimetallic systems. Bragg's law for fcc structures is useful to calculate the lattice constant of the alloyed samples. Datta and Singh [20], in their work on PtRu/C catalyst, used the lattice parameters to calculate the amount of Ru dissolved in the Pt from Vegard's law relationship:

$$a(\text{Pt} - \text{Ru}) = a(\text{Pt}) - kx(\text{Ru}) \qquad (2.1)$$

where x, a (Pt), and a (Pt–M) represent the atomic fraction of the alloyed Ru in the catalyst, lattice parameter of Pt, and a series of Pt–Ru alloyed catalyst, respectively. In a further study, they evaluated lattice parameters and particle sizes and plotted against the atomic % Pt (obtained from EDX) of PtRu/C alloys, as shown in Fig. 2.7.

It has been shown that the catalytic activity of PtRu/C electrodes towards ethanol oxidation is strongly dependent on the Ru content. The onset potential for ethanol oxidation shows distinct shift with varying Ru content, with the most negative potentials reached at 17 atomic % of Ru. The effect of adding Ru on ethanol oxidation is explained by the well-known bifunctional mechanism as expressed

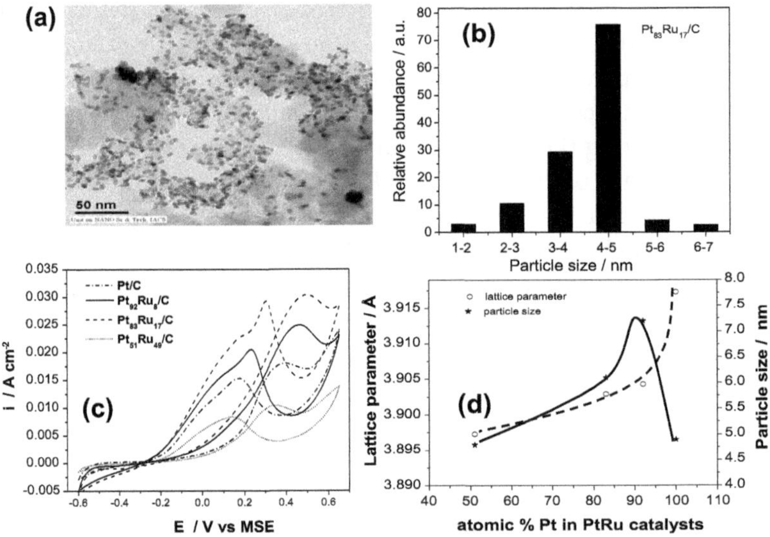

Figure 2.7 (a) TEM images and (b) particle size distribution of Pt/C and PtRu/C electrodes. (c) Cyclic voltammograms in 0.5 mol L^{-1} H_2SO_4 and 1.0 mol L^{-1} EtOH. (d) Lattice parameters and particle size as a function of atomic % Pt in PtRu catalysts.

in the following scheme.

$$\mathrm{Pt + CO \rightarrow Pt - CO_{ad}} \tag{2.2}$$

$$\mathrm{Ru + H_2O \rightarrow Ru - OH_{ad} + H^+ + e^-} \tag{2.3}$$

$$\mathrm{Pt - CO_{ad} + Ru - OH_{ad} \rightarrow CO_2 + H^+ + e^-} \tag{2.4}$$

Datta et al. [46] have also shown that the highest catalytic activity towards ethanol electro-oxidation was also observed for low Ru content (12–17 at%) in the electrodes developed from electrochemical deposition. Typical HRTEM images and SADP pattern for the binary systems are shown in Fig. 2.8.

In view of these observations it was predicted that irrespective of the preparation techniques, 15–20% of Ru in Pt matrix can serve as a better electrocatalyst than Pt itself for EOR in acidic medium. They also explained that Ru does not solely promote ethanol oxidation via the bifunctional mechanism, but the dissociative adsorption of ethanol is also favored by the presence of Ru.

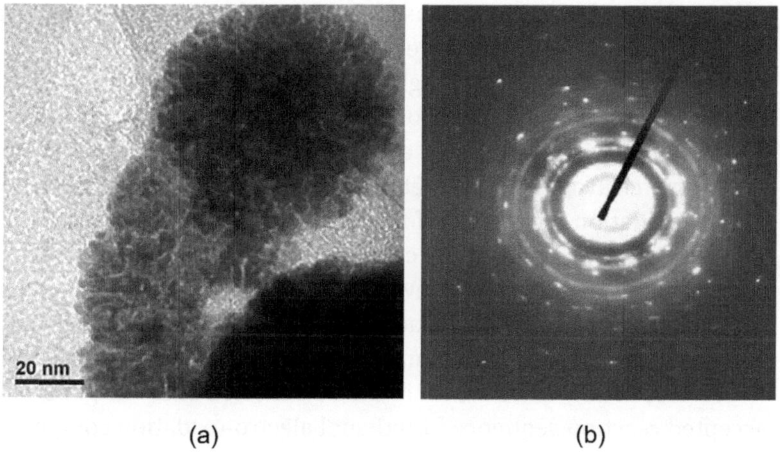

Figure 2.8 (a) HRTEM images of the electrodeposited $Pt_{85}Ru_{15}$ and (b) selected area diffraction pattern (SADP) of the catalyst.

Most importantly, the complete oxidation of ethanol to CO_2 requires the cleavage of the C–C bond, which is between two atoms with little electron affinity or ionization energy, thus making it difficult to break the bond at low temperatures [47]. The partial oxidation of ethanol yields acetaldehyde and acetic acid, which in effect becomes responsible for the decay in electro-efficiency. In recent times, interest has grown in using Rh with Pt because such a combination is likely to be responsible for the C–C bond cleavage during ethanol electro-oxidation.

Tacconi et al. [48] investigated the EOR on Ir and Rh electrodes in acidic media by in situ FTIR technique, and found that the major product with the Ir electrode was acetic acid, but CO_2 with the Rh electrode. De Souza et al. [49] in their study with electrodeposited Pt–Rh layers found that the combination of Pt and Rh in bimetallic electrodes for ethanol oxidation produces a stronger decrease in the acetaldehyde yield compared to pure Pt electrodes, and CO_2 yield is relatively improved. Sheng et al. [50] have shown the role of Rh in $RhPt/CeO_2$ catalyst for efficient dissociation of the C–C bond for ethanol electro-oxidation. The Vesseli group [51] also exhibited the C–C bond cleavage selectivity on single-crystal plane (111) of Rh. Oliveira et al. [52] also demonstrated a large increase in the voltammetric peak current density for the oxidation of small organic molecules at a Pt/Rh bilayer electrodeposited on Pt substrate.

Sengupta et al. and Singh et al. developed PtRh alloyed nanocatalyst respectively through electrochemical and chemical routes and they were able to establish the possible facilitation of C–C bond cleavage by Rh metal in PtRh/C catalysts for ethanol electro-oxidation, and it was found that the application criteria of these materials depend on the chemical composition and crystalline structure of the catalyst developed [42]. They carried out a comparative study on the EOR activity of the electrodeposited Pt and Pt–Rh alloys, and the best ethanol oxidation behavior was observed with introduction of ~30% Rh in the Pt–Rh electrode. The generally accepted reaction sequence for ethanol electro-oxidation comprises the following steps (where M is an active metal site):

Dehydrogenation:

$$M + CH_3 - CH_2OH \rightarrow M - CHOH - CH_3 + H^+ + e^- \quad (2.5.1)$$

$$M - CHOH - CH_3 \rightarrow M - CHO - CH_3 + H^+ + e^- \quad (2.5.2)$$

$$M - CHO - CH_3 \rightarrow M - CO - CH_3 + H^+ + e^- \quad (2.5.3)$$

C–C bond cleavage:

$$M - CO - CH_3 + M \rightarrow M - CO + M - CH_3 \quad (2.6)$$

Water activation:

$$M + H_2O \rightarrow M - OH + H^+ + e^- \quad (2.7)$$

CO_{ads} oxidation:

$$M - CO + M - OH \rightarrow 2M + CO_2 + H^+ + e^- \quad (2.8)$$

Ethanol can also react with adsorbed hydroxyl species directly to produce acetate via a four-electron oxidation pathway:

$$C_2H_5OH + M(OH)_{ads} \rightarrow M(CH_3COO)_{ads} + 4H^+ + 4e^- \quad (2.9)$$

Bergamaski et al. [53] prepared Pt-Rh/C by reduction with formic acid. They studied the influence of the composition of Pt-Rh/C catalysts on the yield of CO_2 using the differential electrochemical mass spectrometry (DEMS) technique and have shown that Pt-Rh/C catalysts enhance the total ethanol oxidation with respect to pure Pt/C by driving the reaction via the CO_2 route. Lima et al. [54]

synthesized carbon-dispersed Pt–Rh (1:1) nanoparticles through chemical reduction method as well as impregnation on carbon powder followed by thermal reduction in hydrogen atmosphere and studied the surface active area by CO stripping and the electrocatalysis of ethanol oxidation by cyclic voltammetry. The Lima and Gonzalez group [55, 56] reported on EOR catalysis with Pt monolayers electrodeposited on carbon-supported Rh nanoparticles. They used a series of sophisticated techniques like in situ XANES (X-ray absorption near-edge structure), FTIR, and others and thereby estimated the CO_2/acetic acid ratio mainly at lower potentials. Colmati et al. [57] synthesized Pt–Rh/C alloy catalysts by reduction of the metal precursors with formic acid and showed the enhanced EOR activity of this catalyst over Pt/C catalyst in acidic medium. Nanoclusters of Pt and Pt–Rh/C were successfully synthesized through polyol method by the Kowal group [58] and superior activity of Pt–Rh/C than Pt/C for EOR was revealed through density functional theory and other experiments. Very recently Shen et al. [59] have shown that owing to the formation of an oxo-metallo-cyclic conformation by microwave polyol process, C–C bond cleavage is the preferential channel for the ethanol dissociation on the Rh surface. In one of the review articles, Antolini [47] has described vividly the work of several research groups on EOR catalysis using multimetallic systems.

Datta and Sengupta [42], in their impedance study on PtRh alloys, have highlighted the enhancement in reaction kinetics with increasing temperature, governed by the partial oxidation of ethanol, which in effect forms more CO_{ads}. In a further study, they have also demonstrated that the impedance data recorded in the complex plane extends to the fourth quadrant at the low frequency end when the temperature is higher than the room temperature (Fig. 2.9) reflecting the inductive behavior of the impedance spectrum. It was considered that CO forms on the electrocatalyst as a strongly adsorbed intermediate and the oxidation of CO_{ads} to CO_2 becomes the rate-determining step. This is all the more likely for the PtRh catalyst where oxygen supply is at premium at a potential of −0.1 V.

In the higher-temperature regime, the charge transfer resistance (R_{ct}) of the best performing catalyst, $Pt_{74}Rh_{26}$ was substantially

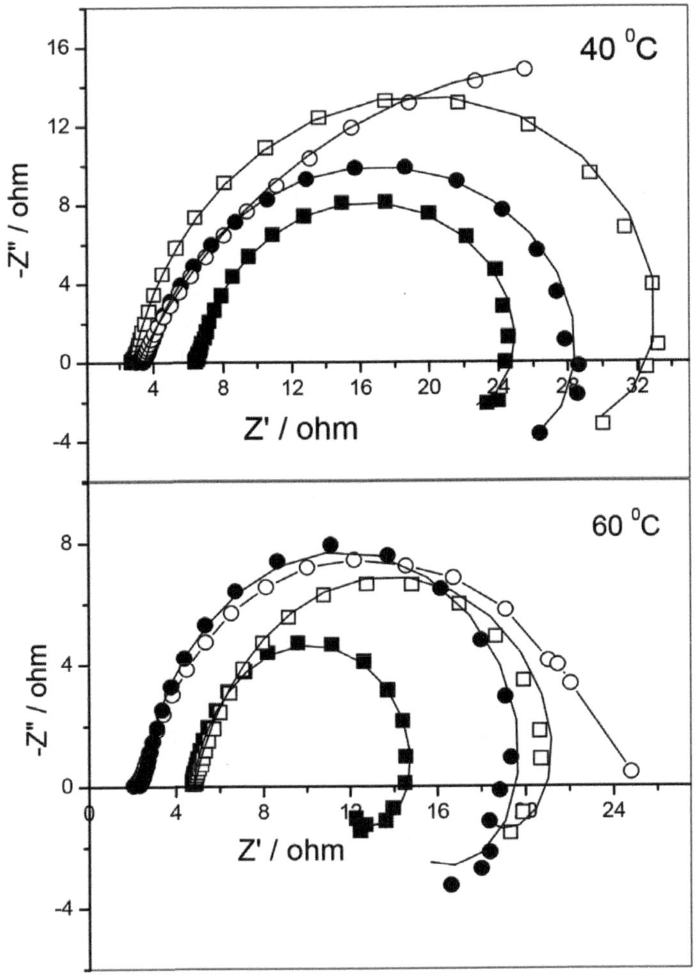

Figure 2.9 Comparison of the Nyquist plots in 0.5 M H_2SO_4+ 1.0 M EtOH at −0.1 V; Pt (•), $Pt_{90}Rh_{10}$ (□), $Pt_{74}Rh_{26}$ (■), and $Pt_{35}Rh_{65}$ (○) electrodes at 40°C and 60°C.

reduced. They suggested that ethanol adsorption can also take place on Rh sites, thereby facilitating a higher degree of participation of Rh as catalyst. Overall, the polarization resistance (R_o) and the inductance (L) values associated with the CO_{ads} oxidation step also

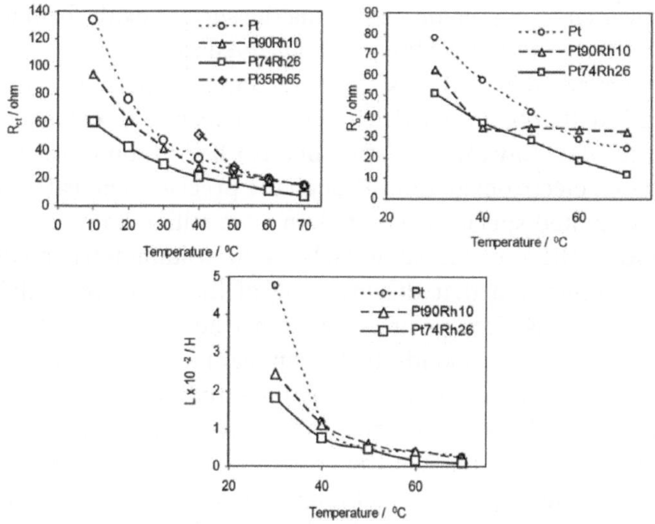

Figure 2.10 Variation of the fitting parameters R_{ct}, R_o, and L with temperature for Pt and PtRh alloy electrodes.

showed a temperature dependence on catalyst composition similar to that observed for the charge transfer resistance (Fig. 2.10).

Investigations have also been made by the same authors on the promotion of electrocatalysis in PtSn nano-catalyst formulation [21, 60]. It has been observed that hydrogen features are somewhat modified in the voltammograms (Fig. 2.11a) of the alloyed catalyst surface. However, the oxidation current is remarkably risen in the

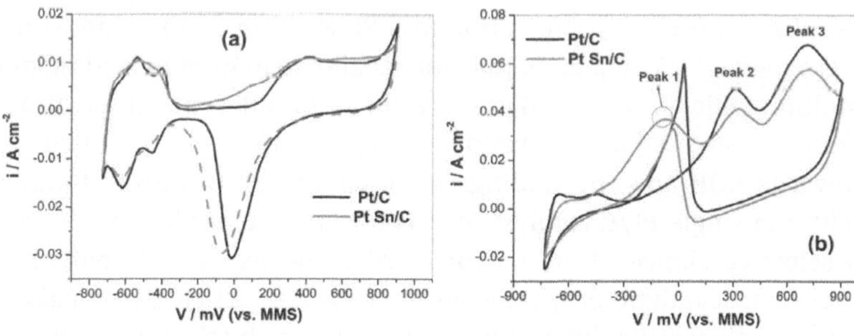

Figure 2.11 Cyclic voltammograms in (a) 0.5 M H_2SO_4 and (b) in solution containing 0.5 M H_2SO_4 and 1.0 M EtOH. Scan rate: 50 mV s^{-1}.

double layer charging region, indicating the surface oxide formation at more negative potentials than that observed for Pt.

They explained that the existence of Sn oxide particles closely associated with the Pt particles seems to excise a catalytically synergistic effect towards oxidation of adsorbed CO on Pt resulting from ethanol electro-oxidation. Though the precise chemical nature of the oxygenated species generated on Sn is still unknown [61], it can be either OH_{ads} or Sn oxide itself. Cyclic voltammetric records clearly demonstrated that the lowering of the onset potential of EOR on a selective composition of PtSn catalysts, by ~300 mV compared to Pt, corresponds to the appearance of an additional anodic peak designated as peak 1 at potentials where Pt surface oxides could not have been formed (Fig. 2.11b). From the results it was predicted that the catalyst formulation with Sn and/or its oxides remaining dispersed in an unalloyed state exhibits the most pronounced enhancement towards ethanol oxidation in comparison to Pt itself.

2.10.2 Ternary Nanocatalysts in Acidic Medium

In course of synthesizing anode materials, attempts are being made to add a third element to the respective binary catalysts. Zhou et al. [62] tested the EOR activity on carbon-supported Pt-based, Pt–Ru–W (1:1:1), and Pt–Ru–Mo (1:1:1) catalyst showed an excellent behavior in acidic medium. Barroso et al. [63] made a comparative study of electrocatalysis on ethanol, bioethanol, and CO oxidation with trimetallic amorphous catalyst having reduced Pt loading. Pt–Ru–Ni/C in (6:3:1) composition and Pt–Ru/C in (1:1) composition were tested by Wang et al. [64] for ethanol oxidation in acid media. Multimetallic alloyed catalysts, PtRuRh/C of different compositions were investigated by Datta and Singh to derive the electro efficiency towards EOR and the findings were compared to those obtained with the single Pt/C catalyst. It was observed that alloying Pt with a selective choice of second or third metal extensively enhances the electrocatalytic properties of the catalyst particles, particularly towards alleviating the poisoning effect caused by the intermediates formed during the course of the reaction sequences and particularly the adsorbed CO-like species. From the study of temperature effect, the activation energy demand was found to be much lower for the optimum composition $Pt_{60}Rh_{20}Ru_{20}/C$ compared to that obtained with Pt alone. Considering different interactions between the ad-atoms influencing the catalytic functionality of the ternary system, the best performance was achieved by the optimum share of partnership between Pt and the cometals in this catalyst where sufficient number of Rh sites are available for C–C bond dissociation subsequent to the dissociative adsorption of ethanol at Pt centers, while the adequate Ru sites facilitates the oxidation of absorbed CO formed as an intermediate at low potentials.

An interesting finding in their extended work on the ternary PtRuRh system is the comparison between MOR (methanol oxidation reaction) and EOR kinetics in acid media, which reflects that the ethanol current density is much higher than that for methanol, indicating that the ternary catalyst is more effective for C_2 molecules. This can be treated as an indirect proof of Rh promoting C–C bond cleavage in ethanol electro-oxidation and also Ru facilitating the oxidation of adsorbed CO.

A careful speculation on their cyclic voltammetric records can justify that Ru has a fair compatibility with the C1 molecule, CH_3OH, while Rh is preferable for the C_2 molecule, ethanol. Figure 2.12a represents the voltammogram for both the alcohols exclusively on the ternary catalyst. The zoomed plot of current density output and the onset potentials as obtained from the voltammograms are presented in Fig. 2.12b. Although the oxidation peak for both the alcohols appears almost at the same potentials, ethanol oxidation is characterized by much higher peak current density, almost 2.5 times the value for methanol. Unlike the binary nanoparticles, the ternary alloy exhibits a high adsorption peak prior to the oxidation, for both the alcohols, however, with much higher peak current for ethanol than for methanol. This is again a reflection of the co-metals of Pt assisting in faster adsorption on the catalyst matrix. The onset of ethanol oxidation commences at a lower potential (−190 mV) than that for methanol (−160 mV), as shown in Fig. 2.12b. Also, when the oxidation currents for both the alcohols are compared at selected potentials, −200 mV and +200 mV, ethanol supersedes methanol with much higher current densities (Fig. 2.12c). Thus, with the help of a second and third metal in addition to Pt, the ternary matrix is

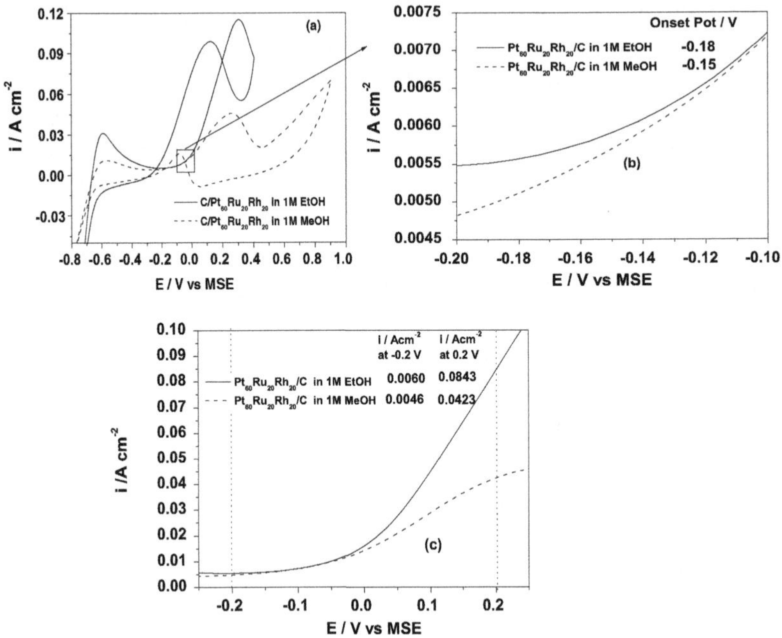

Figure 2.12 CV of $Pt_{60}Ru_{20}Rh_{20}/C$ in 1 mol L^{-1} ethanol and 1 mol L^{-1} methanol; inset (b) onset potentials and (c) current densities at −0.2 V and 0.2 V of both alcohols.

capable of reducing the overvoltage of ethanol electro-oxidation to a considerable extent. The higher degree of catalytic activity observed for the Rh-containing ternary catalyst for ethanol oxidation may be realized in view of its multifaceted capabilities starting from ethanol chemisorption and increased ability to remove protons from the ethanol, which probably facilitates the fragmentation of ethanol molecule, consequently obviating the CO poisoning by complete oxidation of the intermediates to CO_2.

2.10.3 Multimetallic Nanocatalysts for Alkaline DEFC

The attributes of bifunctional mechanism and the electronic effect induced by the electronic interaction of Pt with other metals have so far been discussed, particularly those relevant to the acid environment. However, the selective oxidation of alcohols in alkaline

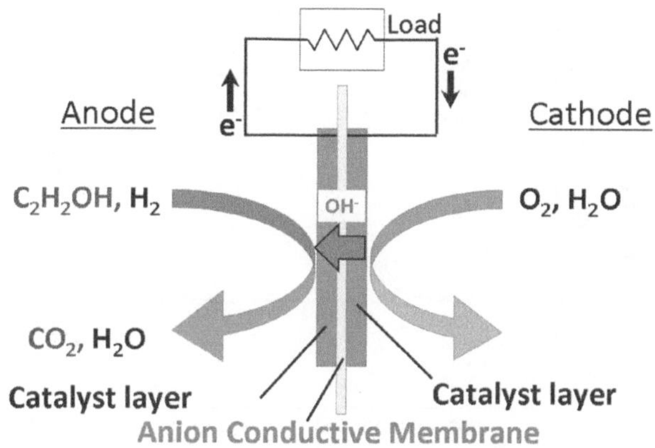

Figure 2.13 Schematic of AEM-FC.

solutions is expected to follow a somewhat different pathway involving a set of different catalysts. Moreover, in recent times with the advent of anion exchange membrane (AEM), there has been upsurge of interest in investigating the catalytic activity of the alloyed/composite materials in order to explore the kinetic advantages of fuel cell reactions in alkaline medium [65]. Figure 2.13 represents the use of AEM in fuel cell. Recent work on the several bimetallic catalysts such as Pt-Ni, Pt-Cr, Pt-Au, and Pt-Co [66, 67] has focused on the understanding of the kinetics and mechanism of ethanol oxidation on these catalysts surface.

A series of oxide-promoted binary catalysts have been reported to show improved activities towards the EOR in alkaline media [68]. Pt alloyed with and/or Ni has given rise to wide attraction in recent years [69]. The Ni-rich subsurface atomic layer is likely to determine the electronic properties of the surface, leading to a reduced coverage of adsorbing species, hydrogen and OH, on the PtNi(111) compared to the pure Pt(111) electrode. The understanding of the stability and atomic structure of the PtNi(111) surface is thus believed to be important in probing the origin of its increased activity for ethanol oxidation [67, 70].

A comprehensive work on the oxidation of methanol, ethanol and 2-propanol has been reported by Datta and coworkers on electrochemically deposited PtPd nanoparticles [26, 71, 72]. In the

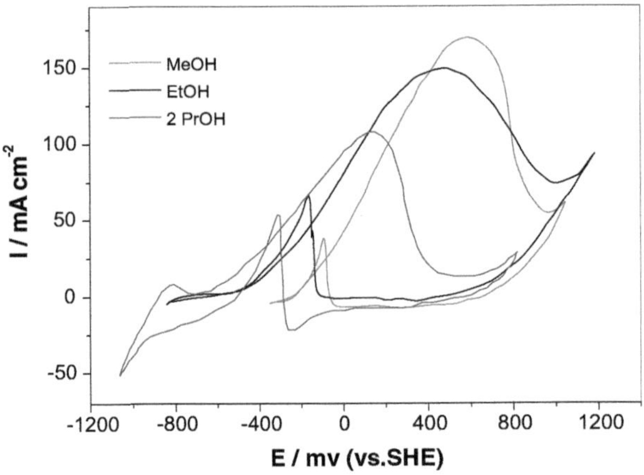

Figure 2.14 Cyclic voltamograms for electro-oxidation of alcohol on the PtPd catalyst in a mixture of 0.5 M NaOH and 1 M alcohol at 40°C.

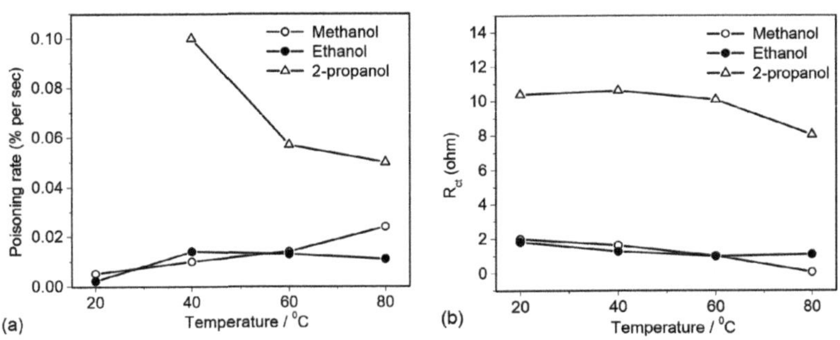

Figure 2.15 Temperature effect on (a) poisoning rate and (b) charge transfer resistance for oxidation of different alcohols at 40 and −160 mV (vs. SHE) on Pt-Pd/C electrode at different temperature.

cyclic voltammetric records (Fig. 2.14), methanol exhibits highest current output. However, the kinetics is compromised by the high overvoltage compared to the other alcohols. The lowest onset potential is observed with 2-propanol. However, with an appreciable current output and a considerably low overvoltage, ethanol appears to be the fuel of merit for direct alcohol fuel cells. This is more discernible in the plots (Fig. 2.15) of temperature effect on the poisoning rate of EOR and the charge transfer resistance determined from the study of impedance. Besides the binary nanocatalysts, there are few reports on effective ternary catalysts for ethanol electro-oxidation for alkaline DEFCs [73–75]. Therefore, a major thrust may be put to investigate the catalytic activity of the multimetallic catalysts in alkaline media for obtaining favorable kinetics and better selectivity toward CO_2 production. Dutta and Datta in their study of electrocatalysis, temperature effect, and reaction mechanism of EOR, have illustrated the role of co-metals Pd and Au in trimetallic PtPdAu matrix [74, 75]. Kinetic control is said to be established on the 1:1:1 combination of this ternary matrix by the oxidation of the carbonaceous intermediates $(CH_3CO)_{ads}$ with OH_{ads}, leading to the formation of CH_3CO^- ions as well as carbonates by facile C–C bond cleavage at a lower overpotential. They explained that these two anionic species exit from the pre-electrode layers and diffuse to the bulk with sufficient ease, thus rendering the surface free from blockage. Eventually, an appreciable reduction in the activation energy, $E_{a(app)}$ of ethanol oxidation, was observed (Fig. 3.16) on this particular composition of the ternary catalyst, which validates its potentiality for alkaline DEFCs.

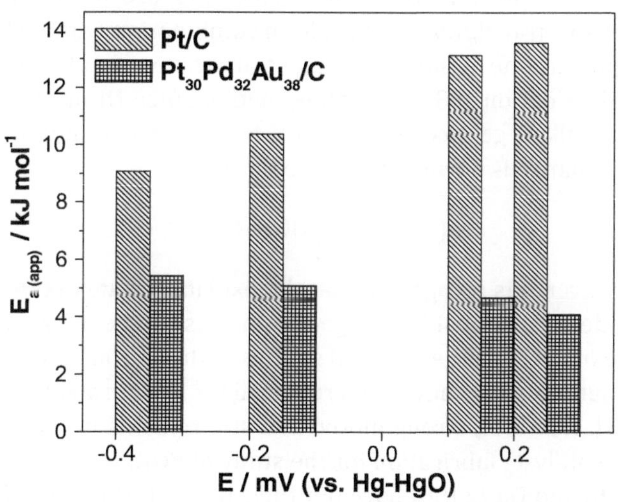

Figure 2.16 Dependence of apparent activation energy with potentials on Pt/C and $Pt_{30}Pd_{38}Au_{32}$/C catalysts in a solution of 0.5 M NaOH and 1.0 M ethanol.

2.10.4 Pt-Free Combinatorial Approach

With regard to the choice of Pt-free electrocatalyst, substantial research efforts have also been devoted to the systematic manipulation of the catalyst composition and structure of Pd-based alloys/composites for alcohol oxidation by the addition of second/third metal oxide promoters [76]. In alkaline media, for instance, Pd and Ru have been investigated for replacing Pt-based catalysts in EOR [77, 78]. These materials have shown marked superiority compared to Pt in terms of reactivity, poison tolerance, and cost. Wang et al. [79] and Xu et al. [80] prepared Pd nanowire arrays by a template electrodeposition method and observed a twofold enhancement of the voltammetric peak current, along with a slow decay of the EOR current in comparison to that shown by the commercial PtRu/C catalysts in alkaline media. Several studies have been reported on Pd-based non-Pt nanocatalysts for use in DEFC. Xu et al. [81] and Hu et al. [82] investigated the activities of oxide-promoted Pd/C electrocatalysts for EOR in alkaline media by voltammetry. K Lee et al. synthesized monodispersed PdNi alloyed nanoparticles for DEFC application [14].

Since the alkaline DAFC also allow the use of comparatively inexpensive transitional metals, Ni in combination with Pd has been attempted as one of the promising Pt-alternative anode catalysts for ethanol oxidation [83] in alkaline media. Since Ni in contact with aqueous alkali gets covered with a layer of nickel hydroxide, the surface change is normally taken to be as

$$\mathrm{Ni(OH)_2} \rightarrow \mathrm{NiOOH} + \mathrm{H^+} + \mathrm{e^-} \tag{2.10}$$

However, it is recognized that the oxidation state of the nickel in the oxide layer probably change continuously between 2 and 4 over a range of potentials. A series of studies with Ni promoted non-noble metal nanocatalysts have been reported for MOR in alkaline solution [84, 85]. However, strong motivation still remains in the pursuit of non-Pt catalysts fabrication for the study of EOR.

Dutta and Datta [86] have put their best effort in identifying the effective formulation of Pt alternatives for ethanol electro-oxidation in alkaline environment. Among the various choice of metallites and alloyed particles, one of their work on developing EOR catalyst of Pd base and incorporating either Au as the noble metal or Ni as the non-noble metal in the matrix is worth mentioning. It has been reported that Au in combination with Pt can act as potential electrocatalyst for CO oxidation in alkaline medium since it can readily activate water-forming Au–OH_{ads} at a comparatively low potential by the following reaction [87, 88]:

$$CO_{ads} + Au-OH_{ads} \rightarrow Au-CO-OH \rightarrow Au-COOH \rightarrow Au + CO_2 + H_2O \quad (2.11)$$

It may be realized from Datta and coworker's study that Pd acts as primary active sites for catalyzing the dehydrogenation of ethanol and CO-like intermediate species could be oxidized by the surface oxide of Au to produce CO_2 or other products, releasing the active sites. Moreover, Pd having less ionization potential (8.34) than Au (9.225) may acquire δ^+ charge by transferring electron to the Au sites and weaken the oxygen bound to the surface, thus making the composite matrix responsible for promoting the dissociative adsorption of ethanol.

In their work on optimizing Ni content in PdNi nanostructured alloyed catalysts, Dutta et al. [89] have drawn the analogy between % alloying and % dispersion and correlated with the catalytic activity of the nanoparticles toward EOR in alkaline medium (Fig. 2.17). The overall impression revealed from the surface structural and morphological information is that the matrix with moderate Ni content between (40~60%) can be grown with fairly good dispersion and in particular the relative abundance of the smaller size particle (2–5) nm is considerably high with Ni content below 60%, leading to better dispersion (Fig. 2.18). However, to achieve a comparatively better state of alloying in the matrix, Ni content has to be raised above 60%. The electrochemical parameters show a close relationship with the extent of alloying in the ad-atom matrix. Although narrow size distribution is obtained with the catalyst with 37% Ni, the catalyst with 63% of Ni exhibit greater extent of alloying and superior electrocatalytic activity amongst all the catalysts studied.

Temperature effect on the rate of EOR was studied for the binary catalyst regime and the product formed during the oxidation

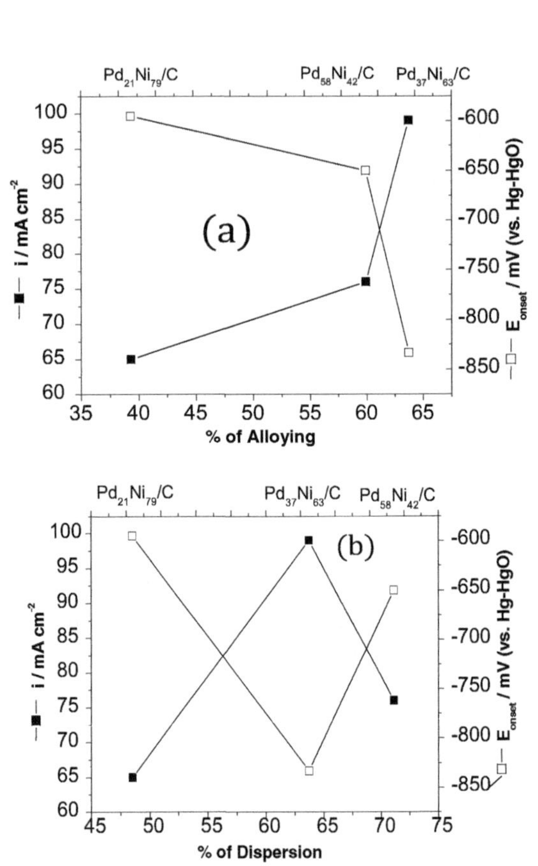

Figure 2.17 Catalytic behavior in terms of oxidation current density and the onset potential as against the (a) % alloying (b) % dispersion.

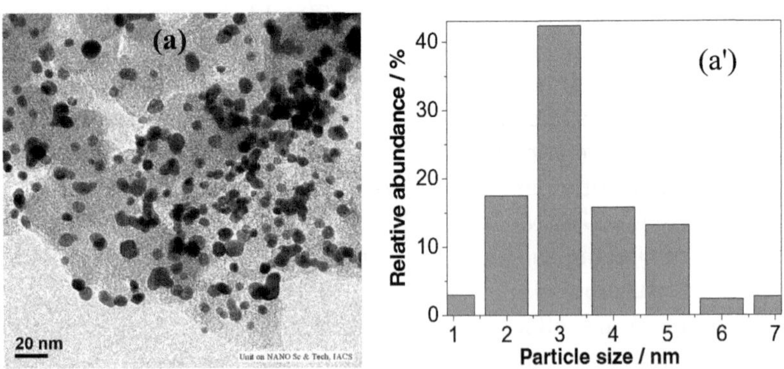

Figure 2.18 TEM images (a) and the particle size distribution (a′) of $Pd_{37}Ni_{63}/C$.

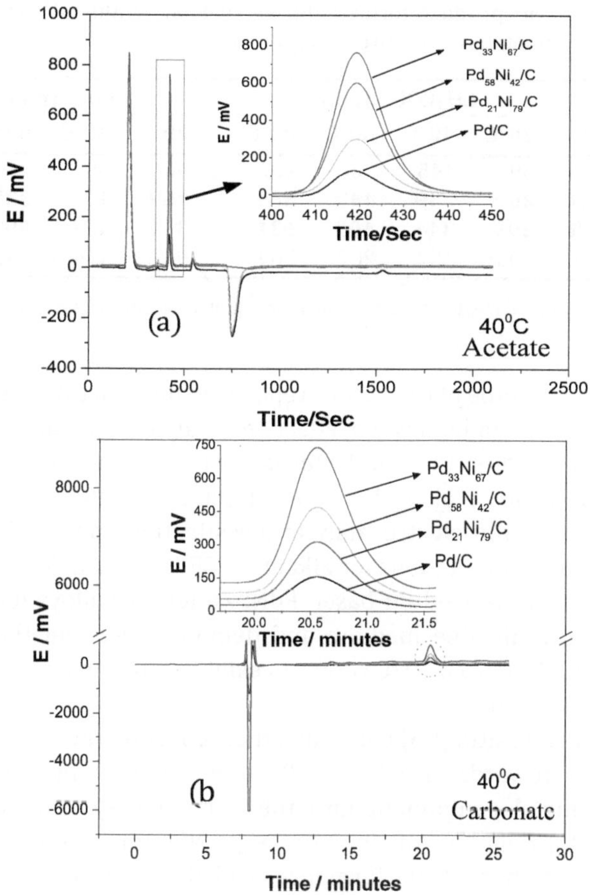

Figure 2.19 Typical Ion chromatograms for (a) acetate and (b) carbonate produced during EOR in a mixture of 0.5 M NaOH and 1.0 M ethanol on Pd/C and PdNi/C electrodes at 40°C. Reproduced with permission from Dutta and Datta [89]. Copyright © 2014 Royal Society of Chemistry.

of ethanol was estimated by ion chromatography, shown in Fig. 2.19. The product yield was remarkably increased, particularly the carbonate, on $Pd_{33}Ni_{67}/C$ when temperature was raised beyond 40°C, as represented in Table 2.1. Alloying at optimized catalyst formulation, therefore, appears to be an important phenomenon that creates favorable reaction centers and help in propagating the EOR to CO_2 production.

Table 2.1 The product formed during EOR at −300 mV on different catalysts estimated by ion chromatography

	CH_3COO^- (ppm)				CO_3^{-2} (ppm)			
Catalysts	**20°C**	**40°C**	**60°C**	**80°C**	**20°C**	**40°C**	**60°C**	**80°C**
Pd/C	89	145	188	207	48	58	76	103
$Pd_{58}Ni_{42}/C$	**209**	**239**	**449**	**589**	**169**	**193**	**338**	**403**
$Pd_{33}Ni_{67}/C$	**293**	**344**	**409**	**631**	**311**	**401**	**597**	**669**
$Pd_{21}Ni_{79}/C$	103	157	207	269	96	134	193	221

Reproduced with permission from Dutta and Datta [89]. Copyright © 2014 Royal Society of Chemistry.

Although some of the earlier reports on fuel cell catalysis include Pt based trimetallic system effective in alkaline media, the work so far reported on the non-Pt ternary catalyst is absolutely scarce. Recently Shen et al. [90] reported on the synthesis of Pt-free catalysts and the electrocatalytic investigations on single, binary, and ternary combinations for alkaline DEFCs. Bambagioni et al. [91] have also synthesized Pd-based PdNi-Zn ternary alloys for ethanol oxidation in alkaline medium and demonstrated that the Pd-(Ni-Zn)/C and Pd-(Ni-Zn-P)/C catalysts exhibited promising activity for ethanol oxidation.

Dutta and Datta [86] continued their effort on the combinatorial approach towards developing Pt free catalysts by introducing a third metal component into the bimetallic system, which led to the formulation of $Pd_{41}Au_{30}Ni_{29}$ NPs on carbon support, exhibiting high-grade performance in alkaline DEFCs. Figure 2.20a,b shows a high-resolution TEM (HRTEM) image of the crystalline matrix composed of the nanostructured $Pd_{41}Au_{30}Ni_{29}/C$ where the trimetallic nanoparticles consist of tightly coupled Au, Pd, and Ni particles, possibly not existing as perfect ternary PdAuNi alloy but rather having obvious interfaces between Au, Pd, and Ni components.

Some well-defined fringe fingerprints were used to measure interplanar d-spacing by line profile, as shown in the insets of Fig. 2.20b, and also with the help of corresponding FFT pattern, shown in the inset. This research group has also shown that addition of Ni to Pd matrix triggers the concurrent promotion of oxidation kinetics due to the existence of NiO/Ni $(OH)_2$ species, on the surface

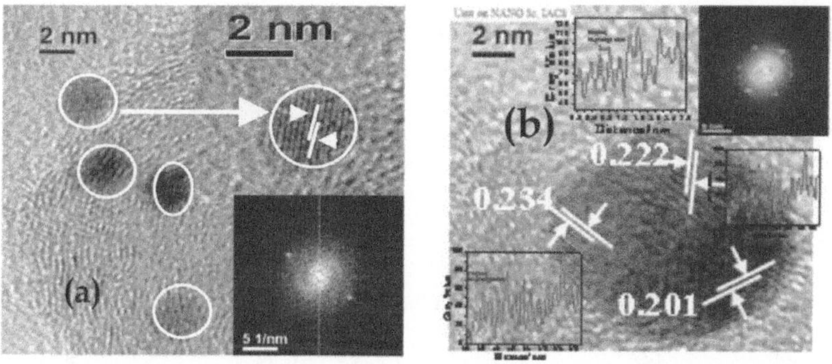

Figure 2.20 (a) HRTEM images and (b) Fringe pattern of a single particle having three different d-spacings. Inset: FFT pattern of corresponding fringe pattern; profile analysis of finger prints for d-spacing evaluation of $Pd_{41}Au_{30}Ni_{29}/C$. Adapted with permission from Dutta and Datta [86]. Copyright © 2012 American Chemical Society.

[86]. Further kinetic input is put forward by the addition of Au in almost the same proportion as Ni to the Pd base metal.

The performances testing with the fabricated anode catalysts in DE(AEM)FC was interesting (Fig. 2.21). The highest performance was shown by the $Pd_{41}Ni_{29}Au_{30}/C$ anode generating an open-circuit potential (OCP) of 0.88 V with a maximum power density output of 51 mW cm^{-2}. The power density generated for the ternary system was found to be almost 38% and 96% greater than the respective binary catalysts PdAu/C and PdNi/C and 168% greater than the single metal catalyst. For the ternary system the discharge curve extends over a longer ohmic region and the maximum current output corresponds to an appreciable amount of ~230 mA cm^{-2}. This reflects a dramatic control of the poisoning effect of the ternary system with respect to the other catalysts.

Dutta et al. [86, 89] also approached for evaluating the direction of charge transfer in between the metal atoms through density functional theory (DFT) computation on the multimetallic catalyst formulation. The DFT method was employed with the Becke's three-parameter hybrid functional (B3LYP), consisting of the nonlocal exchange functional of Becke's three-parameter set and the nonlocal correlation functional of Lee, Yang, and Parr, using the Gaussian 03 package [92, 93]. It transpires from their computation that the

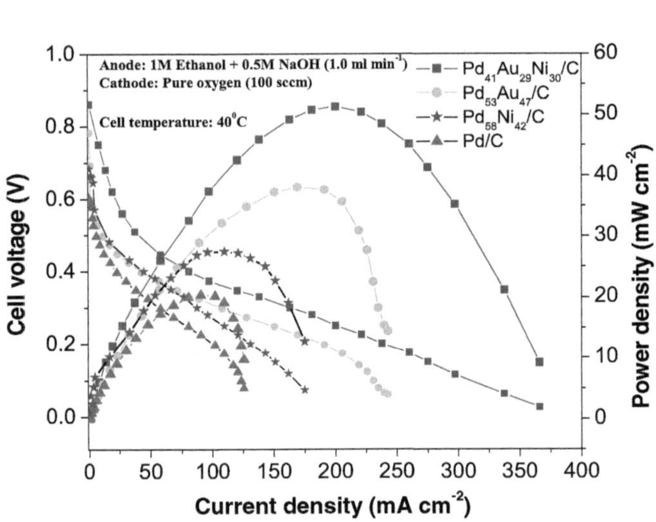

Figure 2.21 Polarization and power density curves of the AEM DEFC with Pd and different binary and ternary anode catalysts (anode: 0.5 M NaOH +1.0 M ethanol). Adapted with permission from Dutta and Datta [86]. Copyright © 2012 American Chemical Society.

geometric size effect of the atomic ensemble is also responsible for the observed changes in the selectivity of the adsorbates and reactivity of the carbonaceous substrates on the surfaces.

Recently, Zhang et al. [94] has reported on the catalytic behavior of AuPtAg ternary catalyst for glucose oxidation. They computed the DFT parameters on the charge transfer process between Ag and Au by taking M55 model consideration on the basis of the particle sizes obtained in the order of 1.5 nm. Dutta et al.'s [86] computation was based on M75 simulated model considering the particle size of the order of 3.0 nm (Fig. 2.22). They explained that the negatively charged gold atoms could activate molecular oxygen by donating an excess electronic charge to the antibonding orbital, and the resulting superoxo- or peroxo-like oxygen may promote ethanol oxidation.

Ni atoms are prone to bear with δ^+ charge by transferring electrons to the neighboring Au and Pd atoms. There may be two modes of electron transfer from Ni to Au directly and via Pd leading to charge distribution among the metallites in the cluster so that Ni carries δ^+ and each of the other two atoms Au and Pd, δ^- charges.

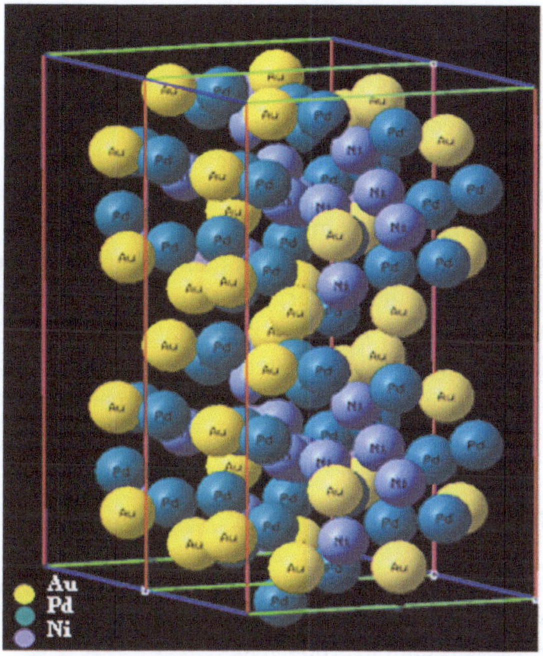

Figure 2.22 Simulated M75 electronic cluster model of $Pd_{53}Au_{20}$ Ni_{27} (yellow, Au; blue, Pd; and light blue, Ni).

These kinds of charge transfer effects might also lead to optimize the % d-band character of the alloyed/composite phases of the catalyst particles responsible for the facile adsorption-desorption kinetics during the process of heterogeneous catalysis. A combined effort of all these properties imparts higher catalytic activity to the $Pd_{53}Au_{20}Ni_{27}$/C compared to the binary PdAu/C and PdNi/C catalysts toward the EOR.

Table 2.2 Mulliken charges of the co-metals in $Pd_{53}Au_{20}Ni_{27}$

Mullikan charges	$Pd_{53}Au_{20}Ni_{27}$/C
Pd (ev)	−0.079
Au (ev)	−0.021
Ni (ev)	0.183

2.11 Conclusion

This chapter provided some of the seminal work on the next generation of electrocatalysts for the anodic reaction in DAFC, particularly focusing on the use of ethanol as green fuel. The discussion is based on the present state-of-the-art DAFC technology, including that collected from the original work carried out by the author's group. In summary, the concerted effort throughout the world to extract the best energy output from the formulation of newer catalysts with Pt-free combinatorial approach is expected to result in further maturity of this technology, making it ready for end users at affordable cost, at the end of the 21st century.

References

1. Rabis, A., Rodriguez, P., and Schmid, T. J. (2012). Electrocatalysis for polymer electrolyte fuel cells: recent achievements and future challenges, *ACS Catalyst*, **2**, 864–890.
2. Larminie, J., and Dicks, A. (2000). Fuel Cell Systems Explained. Bognor Regis: Wiley.
3. Vielstich, W. (1965). *Fuel Cells: Modern Processes for the Electrochemical Production of Energy*. New York: Wiley-Interscience, pp. 374–376.
4. Appleby, A. J., Lloyd, A. C., and Dyer, C. K. (1999). The future of fuel cells, *Scientific American*, **7**, 72.
5. Bossel, U. (2000). *The Birth of the Fuel Cell*, 1835–1845. Oberrohrdorf, Switzerland: European Fuel Cell Forum.
6. Qian, W., Wilkinson, D. P., Shen, J., Wang, H., and Zhang, J. (2006). Architecture for portable direct liquid fuel cells, *J. Power Sources*, **154**, 202–213.
7. Xu, C., Tian, Z., Zhaochen, Z., and Jiang, S. P. (2008). Pd/C promoted by Au for 2-propanol electrooxidation in alkaline media, *Electrochem. Commun.*, **10**, 246–249.
8. Sheikh, A. M., Correa, P. S., Leal da Silva, E., Savaris, I. D., Amico, S. C., and Malfatti, C. F. (2013). Energy conversion using Pd-based catalysts in direct ethanol fuel cell, *Renewable Energy Power Qual. J.*, **11,** ISSN 2172-038 X.

9. Scott, K., Yu, E., Valchogiannopoulos, G., Shivare, M., and Duteanu, N. (2008). Performance of a direct methanol alkaline membrane fuel cell, *J. Power Sources*, **175**, 452–457.

10. Rodrigues, I. A., and Nart, F. C. (2006). 2-Propanol oxidation on platinum and platinum–rhodium electrodeposits, *J. Electroanal. Chem.*, **590**, 145–151.

11. Xu, C., Faghri, A., Li, X., and Ward, T. (2010). Methanol and water crossover in a passive liquid-feed direct methanol fuel cell, *Int. J. Hydrogen Energy*, **35**, 1769–1777.

12. Lamy, C., Coutanceau, C., and Leger, J. M. (2009). The direct ethanol fuel cell: a challenge to convert to bioethanol cleanly into electric energy, in *Catalysis for Sustainable Energy Production*, ed. P. Barbaro and C. Bianchini. Weinheim: Willy-VCH Verlag GmbH.

13. Otomo, J., Kobayashi, X. Li., Wen, T. C., Nagamoto, H., and Takahashi, H. (2004). AC-impedance spectroscopy of anodic reactions with adsorbed intermediates: electro-oxidations of 2-propanol and methanol on carbon-supported Pt catalyst, *J. Electroanal Chem.* **573**, 99–109.

14. Lee, K., Kang, S. W. Lee, S. U., Park, K. H. Lee, Y. W., and Han, S. W. (2012). One-pot synthesis of monodisperse 5 nm Pd–Ni nanoalloys for electrocatalytic ethanol oxidation, *ACS Appl. Mater. Interfaces*, **4**, 4208–4214.

15. Kowal, A., Gojkovic, S. L. J., Lee, K. S., Olszewski, P., and Sung, Y. E. (2009). Synthesis, characterization and electrocatalytic activity for ethanol oxidation of carbon supported Pt, Pt–Rh, Pt–SnO2 and Pt–Rh–SnO2 nanoclusters, *Electrochem Comm*, **11**, 724–727.

16. Lamy, C., Belgsir, E. M., and Leger, J. M. (2001). Electrocatalytic oxidation of aliphatic alcohols: application to the direct alcohol fuel cell (DAFC), *J. Appl. Electrochem.*, **31**, 799–809.

17. Kim, I., Bong, S., Woo, S., Mahajan, R. K., and Kim, H. (2011). Highly active 40 wt.% PtRu/C anode electrocatalysts for PEMFCs prepared by an improved impregnation method, *Int. J. Hydrogen Energy*, **36**, 1803–1812.

18. Velázquez-Palenzuela, A., Centellas, F., Garrido, J. A., Arias, C., Rodríguez, R. M., Brillas, E., and Cabot, P. L. (2011). Kinetic analysis of carbon monoxide and methanol oxidation on high performance carbon-supported Pt–Ru electrocatalyst for direct methanol fuel cells, *J. Power Sources*, **196**, 3503–3512.

19. Tsivadze, A. Yu., Tarasevich, M. R., Andreev, V. N., and Bogdanovskaya, V. A. (2007). Prospects of low-temperature platinum-free fuel cells, *Russ. J. Gen. Chem.*, **77**, 783–789.

20. Datta, J., Singh, S., Das, S., and Bandhyopadhyay, N. R. (2009). A comprehensive study on the effect of Ru addition to Pt electrodes for direct ethanol fuel cell, *Bull. Mater. Sci.*, **32**, 643–652.
21. Sen Gupta, S., Singh, S., and Datta, J. (2010). Temperature effect on the electrode kinetics of ethanol electro-oxidation on Sn modified Pt catalyst through voltammetry and impedance spectroscopy, *Mat. Chem. Phys.* **120**, 682–690.
22. Sen Gupta, S., Mahapatra, S. S., and Datta, J. (2004). A potential anode material for the direct alcohol fuel cell, *J. Power Sources,* **131**, 169–174
23. Shen, P. K., and Xu, C. (2006). Alcohol oxidation on nanocrystalline oxide Pd/C promoted electrocatalysts, *Electrochem. Commun.*, **8**, 184–188.
24. Hu, F., Chen, C., Wang, Z., Wei, G., and Shen, P. K. (2006). Mechanistic study of ethanol oxidation on Pd–NiO/C electrocatalyst, *Electrochim. Acta*, **52**, 1087–1091.
25. Bagchi, J., and Bhattacharya, S. K. (2007). Electrocatalytic activity of binary Palladium Ruthenium anode catalyst on Ni-support for ethanol alkaline fuel cells, *Trans. Met. Chem.*, **32**, 47–55.
26. Sengupta, S., and Datta, J. (2005). An investigation into the electro-oxidation of ethanol and 2-propanol for application in direct alcohol fuel cells (DAFCs), *J. Chem. Sci.*, **117**, 337–344.
27. Liu, J., Li, Z., He, C., Fu, R., Wu, D., and Song, S. (2011). Wormholelike mesoporous carbons as the support for Pt_2Sn_1 towards ethanol electrooxidation: Effect of pore diameter, *Int. J. Hydrogen Energy*, **36**, 2250–2257.
28. Sevjidsuren, G., Zils, S., Kaserer, S., Wolz, A., Ettingshausen, F., Dixon, D., Schoekel, A., Roth, C., Altantsog, P., Sangaa, D., and Ganzorig, Ch. (2010). Effect of different support morphologies and pt particle sizes in electrocatalysts for fuel cell applications, *J. Nanomater.*, **2010**, 1–9.
29. Ettingshausen, F., Weidner, A., Zils, S., Wolz, A., Suffner, J., Michel, M., and Roth, C. (2009). Alternative support materials for fuel cell catalysts, *ECS Trans.*, **25**, 1883–1892.
30. Datta, J., Dutta, A., and Biswas, M. (2012). Enhancement of functional properties of PtPd nano catalyst in metal-polymer composite matrix: Application in direct ethanol fuel cell, *Electrochem. Commun.*, **20**, 56–59.
31. Rajalakshmi, N., and Dhathathreyan, K. S. (2008). Nanostructured platinum catalyst layer prepared by pulsed electrodeposition for use in PEM fuel cells, *Int. J. Hydrogen Energy*, **33**, 5672–5677.

32. Liu, Z., Lee, J., Han, M., Chen, W., and Gan, L. (2002). Synthesis and characterization of PtRu/C catalysts from microemulsions and emulsions, *J. Mater. Chem.*, **12**, 2453–2458.

33. Sakthivel, M., Schlange, A., Kunz, U., and Turek, T. (2010). Microwave assisted synthesis of surfactant stabilized platinum/carbon nanotube electrocatalysts for direct methanol fuel cell applications, *J. Power Sources*, **195**, 7083–7089.

34. Hui, C. L., Li, X. G., and Hsing, I. M. (2005). Well-dispersed surfactant-stabilized Pt/C nanocatalysts for fuel cell application: dispersion control and surfactant removal, *Electrochim. Acta*, **51**, 711–719.

35. Lim, D. H., Lee, W. D., Choi, D. H., Park, D. R., and Lee, H. I. (2008). Preparation of platinum nanoparticles on carbon black with mixed binary surfactants: Characterization and evaluation as anode catalyst for low-temperature fuel cell, *J. Power Sources*, **185**, 159–165.

36. Singh, S., and Datta, J. (2010). Size control of Pt nanoparticles with stabilizing agent for better utilization of the catalyst in fuel cell reaction, *J. Mater. Sci.*, **45**, 3030–3040.

37. Pan, C., Dassenoy, F., Casanove, M. J., Philippot, K., Amiens, C., Lecante, P., Mosset, A., and Chaudret, B. (1999). A New Synthetic Method toward Bimetallic Ruthenium Platinum Nanoparticles; Composition Induced Structural Changes, *J. Phys. Chem. B*, **103**, 10098–10101.

38. Witham, C. K., Chun, W., Valdez, T. I., and Narayanan, S. R. (2000). Performance of direct methanol fuel cells with sputter-deposited anode catalyst layers, *Electrochem. Solid-State Lett.*, **3**, 497–500.

39. Zhanga, X., Donga, D., Li, D., Williams, T., Wanga, H., Webleya, P. A. (2009). Direct electrodeposition of Pt nanotube arrays and their enhanced electrocatalytic activities, *Electrochem. Comns.*, **11**, 190–193.

40. Stab, G., and Urban, P. (2001). Process for the Manufacture of an Electrode for a Solid Polymer Fuel Cell, *US Patent 6*, **258**, 239.

41. Sen Gupta, S., and Datta, J. (2005). Electrode kinetics of ethanol oxidation on novel CuNi alloy supported catalysts synthesized from PTFE suspension, *J. Power Sources*, **145**, 124–132.

42. Sen Gupta, S., and Datta, J. (2006). A comparative study on ethanol oxidation behavior at Pt and PtRh electrodeposits, *J. Electroanal. Chem.*, **594**, 65–72.

43. Dutta, A., Mahapatra, S. S., and Datta, J. (2011). High performance PtPdAu nano-catalyst for ethanol oxidation in alkaline media for fuel cell applications, *Int. J. Hydrogen Energy*, **36**, 14898–14906.

44. Kakati, N. Maiti, J., Jee, S. H., Lee, S. H., and Yoon, Y. S. (2011). Carbon nanotube–chitosan composite electrodes for electrochemical removal of Cu(II) ions, *J. Alloys Compd.*, **509**, 5667–5617.

45. Kim, I., Bong, S., Woo, S., Mahajan, R. K., and Kim, H. (2011). Highly active 40 wt.% PtRu/C anode electrocatalysts for PEMFCs prepared by an improved impregnation method, *Int. J. Hydrogen Energy*, **36**, 1803–1812.

46. Datta, J., Sen Gupta, Singha, S. S., Mukherjee, S., and Mukherjee, M. (2011). Significant role of Ru-oxide present in the Pt-Ru alloy catalyst for ethanol electro-oxidation in acid medium, *Mater. Manuf. Processes*, **26**, 261–271.

47. Antolini, E., and Gonzalez, E. R. (2010). A simple model to assess the contribution of alloyed and non-alloyed platinum and tin to the ethanol oxidation reaction on Pt–Sn/C catalysts: application to direct ethanol fuel cell performance, *Electrochim. Acta*, **55**, 6485–6490.

48. Tacconi, N. R., Lezna, R. O., Beden, B., Hahn, F., and Lamy, C. (1994). In-situ FTIR study of the electrocatalytic oxidation of ethanol at iridium and rhodium electrodes, *J. Electroanal. Chem.*, **379**, 329–337.

49. de Souza, J. P. I., Queiroz, S. L., Bergamaski, K., Gonzales, E. R., and Nart, F. C. (2002). Electro-oxidation of ethanol on Pt, Rh, and PtRh electrodes. A study using DEMS and in-situ FTIR techniques, *J. Phys. Chem. B*, **106**, 9825–9830.

50. Sheng, P. Y., Yee, A., Bowmaker, G. A., and Idriss, H. (2002). H_2 production from ethanol over Rh–Pt/CeO_2 catalysts: the role of Rh for the efficient dissociation of the carbon–carbon bond, *J. Catal.*, **208**, 393–403.

51. Vesselli, E., Baraldi, A., Comelli, G., Lizzit, S., and Rosei, R., (2004). Ethanol decomposition: C–C cleavage selectivity on Rh(111), *Chem. Phys. Chem.*, **5**, 1133–1140.

52. Oliveira, R. T. S., Santos, M. C., Marcussi, B. G., Nascente, P. A. P., Bulhões, L. O. S., and Pereira, E. C. (2005). The use of a metallic bilayer for the oxidation of small organic molecules, *J. Electroanal. Chem.*, **575**, 177–182.

53. Bergamaski, K., Gonzales, E. R., and Nart, F. C. (2008). Ethanol oxidation on carbon supported platinum-rhodium bimetallic catalysts, *Electrochim. Acta*, **53**, 4396–4406.

54. Lima, F. H. B., Profeti, D., Valbuena, W. H. L., Ticianelli, E. A., and Gonzalez, E. R. (2008). Carbon-dispersed Pt–Rh nanoparticles for ethanol electro-oxidation. Effect of the crystallite size and of temperature, *J. Electroanal. Chem.*, **617**, 121–129.

55. Lima, F. H. B., and Gonzalez, E. R. (2008). Electrocatalysis of ethanol oxidation on Pt monolayers deposited on carbon-supported Ru and Rh nanoparticles, *Appl. Catal. B: Environ.*, **79**, 341–346.

56. Lima, F. H. B., and Gonzalez, E. R. (2008). Ethanol electro-oxidation on carbon-supported Pt–Ru, Pt–Rh and Pt–Ru–Rh nanoparticles, *Electrochim. Acta,* **53**, 2963–2971.

57. Colmati, F., Antolini, E., and Gonzalez, E. R. (2008). Preparation, structural characterization and activity for ethanol oxidation of carbon supported ternary Pt–Sn–Rh catalysts, *J. Alloys Compd.*, **456**, 264–270.

58. Kowal, A., Li, M., Shao, M., Sasaki, K., Vukmirovic, M. B., and Zhang, J. (2009). Ternary Pt/Rh/SnO_2 electrocatalysts for oxidizing ethanol to CO, *Nat. Mater.*, **8**, 325–330.

59. Shen, S. Y., Zhao, T. S., and Xu, J. B. (2010). Carbon supported PtRh catalysts for ethanol oxidation in alkaline direct ethanol fuel cell, *Int. J. Hydrogen Energy,* **35**, 12911–12917.

60. Sen Gupta, S., Singh, S., and Datta, J. (2009). Promoting role of unalloyed Sn in PtSn binary catalysts for ethanol electro-oxidation, *Mater. Chem. Phys.,* **116**, 223–228.

61. Zignani, S. C., Gonzalez, E. R., Baglio, V., Siracusano, S., Aricò, A. S. (2012). Investigation of a Pt_3Sn/C electro-catalyst in a direct ethanol fuel cell operating at low temperatures for portable applications, *Int. J. Electrochem. Sci.*, **7**, 3155–3166.

62. Zhou, W., Zhou, Z., Song, S., Li, W., Sun, G., Tsiakaras, P., and Xin, Q. (2003). Pt based anode catalysts for direct ethanol fuel cells, *Appl. Catal. B,* **46**, 273–285.

63. Barroso, J., Pierna, A. R., Blanco, T. C., and Ruiz, N. (2014). Trimetallic amorphous catalyst with low amount of platinum: Comparative study for ethanol, bioethanol and CO electrooxidation, *Int. J. Hydrogen Energy,* **39**, 3984–3990.

64. Wang, Z., Yin, G., Zhang, J., Sun, Y., and Shi, P. (2006). Co-catalytic effect of Ni in the methanol electro-oxidation on Pt–Ru/C catalyst for direct methanol fuel cell, *Electrochim. Acta,* **51**, 5691–5697.

65. Singh, R. N., and Singh, A. (2009). Anindita Electrocatalytic activity of binary and ternary composite films of Pd, MWCNT and Ni, Part II: methanol electrooxidation in 1 M KOH, *Int. J. Hydrogen Energy,* **34**, 2052–2057.

66. Ko, E. Y., Park, E. D., Seo, K. W., Lee, H. C., Lee, D., and Kim, S. (2006). Pt–Ni/γ-Al_2O_3 catalyst for the preferential CO oxidation in the hydrogen stream, *Catal. Lett.*, **110**, 275–279.

67. Mania, P., Srivastavaa, R., and Strasser, P. (2011). Dealloyed binary PtM3 (M = Cu, Co, Ni) and ternary PtNi3M (M = Cu, Co, Fe, Cr) electrocatalysts for the oxygen reduction reaction: Performance in polymer electrolyte membrane fuel cells, *J. Power Sources,* **196**, 666–673.

68. He, Q., Chen, W., Mukerjee, S., Chen, S., and Laufek, F. (2009). Carbon-supported PdM (M = Au and Sn) nanocatalysts for the electrooxidation of ethanol in high pH media, *J. Power Sources*, **187**, 298–304.
69. Yang, H., Coutanceau, C., Leger, J. M., Alonso-Vante, N., and Lamy, C. (2005). Methanol tolerant oxygen reduction on carbon-supported Pt–Ni alloy nanoparticles, *J. Electroanal. Chem.*, **576**, 305–313.
70. Hoover, R. R., and Tolmachev, Y. V. (2009). Electrochemical properties of Pt coatings on Ni prepared by atomic layer deposition, *J. Electrochem. Soc.*, **156**, A37–A43.
71. Mahapatra, S. S., and Datta, J. (2011). Characterization of Pt-Pd/C electrocatalyst for methanol oxidation in alkaline medium, *Int. J. Electrochem.*, **2011**, 1–16.
72. Mahapatra, S. S., Dutta, A., and Datta, J. (2010). Temperature effect on the electrode kinetics of ethanol oxidation on Pd modified Pt electrodes and the estimation of intermediates formed in alkali medium, *Electrochim. Acta*, **55**, 9097–9104.
73. Silva, D., Geraldes, A., Cardoso, E., Tusi, M., Linardi, M., Spinace, E., and Neto, A. (2011). Preparation of PtAu/C and PtAuBi/C electrocatalysts using electron beam irradiation for methanol and ethanol electro-oxidation in alkaline medium, *Int. J. Electrochem. Sci.*, **6**, 3594–3606.
74. Datta, J., Dutta, A., and Mukherjee, S. (2011). The beneficial Role of the cometals Pd and Au in the carbon-supported PtPdAu catalyst toward promoting ethanol oxidation kinetics in alkaline fuel cells: temperature effect and reaction mechanism, *J. Phys. Chem. C*, **115**, 15324–15334.
75. Dutta, A., Mahapatra, S., and Datta, J. (2011). High performance PtPdAu nano-catalyst for ethanol oxidation in alkaline media for fuel cell applications, *Int. J. Hydrogen Energy*, **36**, 14898–14906.
76. Wang, M., Guo, D. J., and Li, H. L. (2005). High activity of novel Pd/TiO_2 nanotube catalysts for methanol electro-oxidation, *J. Solid State Chem.*, **178**, 1996–2000.
77. Xu, C., Cheng, L., Shen, P., and Liu, Y. (2007). Methanol and ethanol electrooxidation on Pt and Pd supported on carbon microspheres in alkaline media, *Electrochem. Commun.*, **9**, 997–1001.
78. Hu, F., Ding, F., Song, S., and Shen, P. (2006). Pd electrocatalyst supported on carbonized TiO_2 nanotube for ethanol oxidation, *J. Power Sources*, **163**, 415–419.
79. Wang, H., Xu, C., Cheng, F., and Jiang, S. (2007). Pd nanowire arrays as electrocatalysts for ethanol electrooxidation, *Electrochem. Commun.*, **9**, 1212–1216.

80. Xu, C., Wang, H., Shen, P., and Jiang, S. (2007). Highly ordered Pd nanowire arrays as effective electrocatalysts for ethanol oxidation in direct alcohol fuel cells, *Adv. Mater.*, **19**, 4256–4259.

81. Xu, C., Tian, Z., Shen, P., and Jiang, S. P. (2008). Oxide (CeO_2, NiO, Co3O4 and Mn_3O_4)$^-$ promoted Pd/C electrocatalysts for alcohol electrooxidation in alkaline media, *Electrochim. Acta*, **53**, 2610–2618.

82. Hu, F., Chen, C., Wang, Z., Wei, G., and Shen, P. K. (2006). Mechanistic study of ethanol oxidation on Pd–NiO/C electrocatalyst, *Electrochem. Acta*, **52**, 1087–1091.

83. Shen, S. Y., Zhao, T. S., Xu, J. B., and Li, Y. S. (2010). Synthesis of PdNi catalysts for the oxidation of ethanol in alkaline direct ethanol fuel cells, *J. Power Sources*, **195**, 1001–1006.

84. Shobha, T., Aravinda, C. L., Bera, P., Gomathi Devi, L., and Mayanna, S. M. (2003). Characterization of Ni–Pd alloy as anode for methanol oxidative fuel cell, *Mater. Chem. Phys.*, **80**, 656–661.

85. Rahim, M. A. A., Hameed, R. M. A., and Khalil, M. W. (2004). Nickel as a catalyst for the electro-oxidation of methanol in alkaline medium, *J. Power Sources*, **134**, 160–169.

86. Dutta, A., and Datta, J. (2012). Outstanding catalyst performance of PdAuNi nanoparticles for the anodic reaction in an alkaline direct ethanol (with anion-exchange membrane) fuel cell, *J. Phys. Chem. C*, **116**, 25677–25688.

87. Wu, Z. L., Zhou, S. H., Zhu, H. G., Dai, S., and Overbury, S. H. (2008). Oxygen-assisted reduction of Au species on Au/SiO_2 catalyst in room temperature CO oxidation, *Chem. Commun.*, 3308–3310.

88. Chenga, F., Daia, X., Wanga, H., Jiangb, S., Zhanga, M., and Xu, C. (2010). Synergistic effect of Pd–Au bimetallic surfaces in Au-covered Pd nanowires studied for ethanol oxidation, *Electrochim. Acta*, **55**, 2295–2298.

89. Dutta, A., and Datta, J. (2014). Energy efficient role of Ni/NiO in PdNi nano catalyst used in alkaline DEFC, *J. Mater. Chem. A*, **2**, 3237–3250.

90. Shen, S., Zhao, T. S., Xu, J., and Li, Y. (2011). High performance of a carbon supported ternary PdIrNi catalyst for ethanol electro-oxidation in anion-exchange membrane direct ethanol fuel cells, *Energy Environ. Sci.*, **4**, 1428–1433.

91. Bambagioni, V., Bianchini, C., Filippi, J., Oberhauserlal, W., Marchionni, A., Vizza, F., Psaro, R., Sordelli, L., Foresti, M. L., and Innocenti, M. (2009). Ethanol oxidation on electrocatalysts obtained by spontaneous deposition of palladium onto nickel-zinc materials, *ChemSusChem*, **2**, 99–112.

92. Becke, A. D. (1993). Density-functional thermochemistry. III. The role of exact exchange, *J. Chem. Phys.*, **98**, 5648–5652.

93. Reed, A. E., Curtiss, L. A., and Weinhold, F. (1988). Intermolecular interactions from a natural bond orbital, donor-acceptor viewpoint, *Chem. Rev.*, **88**, 899–926.

94. Zhang, H., Okumura, M., and Toshima, N. (2011). Stable dispersions of PVP-protected Au/Pt/Ag trimetallic nanoparticles as highly active colloidal catalysts for aerobic glucose oxidation, *J. Phys. Chem. C*, **115**, 14883–14891.

Chapter 3

Understanding Electrocatalytic Activity Enhancement of Bimetallic Nanoparticles to Ethanol Electro-oxidation Reaction

Yixuan Wang and Zhenfeng Xu

Department of Natural Science, Albany State University, Albany, Georgia 31705, USA
yixuan.wang@asurams.edu

This chapter summarizes the systematic density functional theory studies on the adsorptions and decompositions of water and ethanol on Pt_nM ($n = 2, 3, 6$, and 9; M = Pt, Sn, Ru, Rh, Pd, Cu, and Re). The results show that from both kinetic and thermodynamic viewpoints Sn is more active to water decomposition than pure Pt and all of the other investigated PtM except for PtRe, which well supports the assumption of the bifunctional mechanism that Sn site accelerates the dissociation of H_2O. *However, although the energy barrier for water decomposition on the Re site of the PtRe is only slightly lower the that on Sn site, the adsorption energy and dissociation energy on the Re site are considerably more favorable than those on the Sn site of PtS.* Although the α-hydrogen adsorption is weaker than the hydroxyl of ethanol, the adsorptions of ethanol on Pt and M

Nanomaterials for Direct Alcohol Fuel Cell
Edited by Yixuan Wang

ISBN 978-981-4669-00-9 (Hardcover), 978-981-4669-01-6 (eBook)
www.panstanford.com

sites of PtM via the α-hydrogen were well established, and the potential energy profiles show that the dehydrogenation via the α-H path has much lower energy barrier than that via the hydroxyl path. Generally for the α-hydrogen path the adsorption is a rate-determining step because of rather low dehydrogenation barrier for the α-hydrogen adsorption complex (*thermodynamic control*), while the hydroxyl path is determined by its dehydrogenation step (*kinetic control*). Alloying of Sn enhances the adsorption of ethanol on Pt site through the α-H, as well as those on both Pt and Sn sites through the hydroxyl. However, the Re alloying does not enhance the former adsorption although it enhances the latter one on both Pt and Re sites.

3.1 Introduction

A direct ethanol proton exchange membrane fuel cell (PEMFC) offers a promising alternative power source because of its superiority to the most studied H_2-O_2 PEMFC in a few aspects, such as storage and high energy density. The most important is that ethanol is a renewable fuel since it can be produced in a great quantity from biomass and it is much less toxic than methanol as well. However, the complete oxidation of ethanol to H_2O and CO_2 is much more difficult than those of H_2 as well as methanol because of the C–C bond in ethanol. Pt was recognized to be the most active catalyst for the ethanol electro-oxidation reaction (EER), but it is readily poisoned by adsorbed CO species. Much effort has been put into searching for more efficient electrocatalysts for the EER and considerable progresses have also been made [1–6]. For instance, PtSn/C has been demonstrated the ability to enhance the electrocatalytic activity of Pt towards the EER with respect to increasing current density as well as to lowering the onset potential of ethanol oxidation by approximately 0.2 V compared to Pt alone [2, 3, 7, 8]. However, the experimental result shows that ethanol is still far from complete oxidation even at the PtSn/C anode, and the oxidation products are measured to be acetaldehyde (22.8%), acetic acid (74.7%), and CO_2 (2.5%) [9].

The mechanism of ethanol electro-oxidation on the PtSn was widely interpreted with bifunctional mechanism and ligand effect [10, 11], where dissociative adsorption of ethanol occurs only on platinum sites and tin primarily promotes adsorption and dissociation of water to form OH oxidizing intermediates from ethanol [4]. Other researchers suggested rather similar mechanisms for the EER [1, 2, 8, 12]. Although some interesting assumptions have been made, several essential issues still remain uncertain. For example, although Sn and Ru have similar activity for water dissociative adsorption, PtRu does not show as active as PtSn to the EER. Thus, Sn must have another effect besides ligand effect and enhancing H_2O dissociation because otherwise Ru should display almost the same activity as Sn for the EER [4].

To well understand electrocatalytic enhancement of PtSn for the EER and eventually design more efficient catalysts for the EER, it is essential to investigate adsorption and dissociation of both water and ethanol over Pt-based bimetallic particles at an atomic level. Recently, the authors extensively investigated the adsorption and decomposition of water on Pt-based bimetallic clusters with the density functional theory (DFT) [13], and together with other pioneering work concluded that from both kinetic and thermodynamic viewpoints Sn is more active to water decomposition than pure Pt and the PtRu alloy, which well supports the assumption of the bifunctional mechanism that Sn site primarily accelerates the dissociation of H_2O. Regarding the electro-oxidation of ethanol on Pt, much theoretical effort has already been devoted to understanding its reaction mechanism, mainly with plane-wave based DFT-slab calculations [14–16]. However, few theoretical studies have been reported on the EER over the Pt-based bimetallic surface [15, 17]. Alcala et al. discussed the electrocatalytic mechanism of ethanol on Pt/SiO_2 and Pt_3Sn/SiO_2 with periodic, self-consistent DFT with Perdew-Wang 1991 generalized gradient approximation [15], indicating that the lowest energy pathway for acetaldehyde formation is via the ethoxy species ($CH_3CH_2O^*$, reaction a) on Pt_3Sn while via the 1-hydroxyethyl species (CH_3C^*HOH, reaction b; *stands for the adsorption site.) on Pt. Both the α-hydrogen and hydroxyl adsorptions on Pt, as well as on the alloyed transition metal M sites of Pt_3M, were considered as initial reaction steps [17].

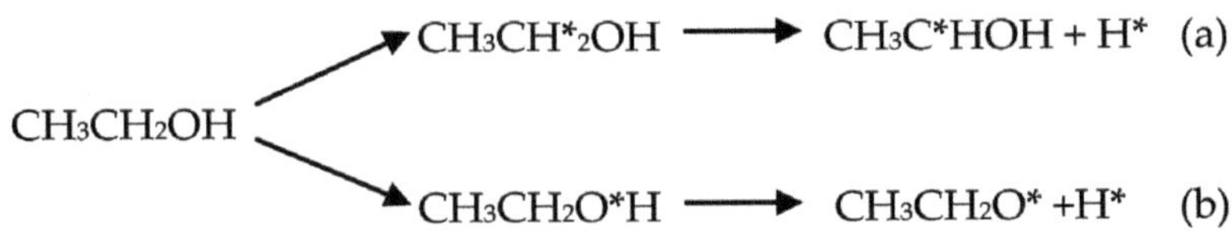

In this chapter the new understanding about the EER will be demonstrated including the detailed adsorptions and decompositions of water and ethanol on a variety of the bimetallic nanoclusters. To clarify the initial steps of the ethanol electro-oxidation reaction on Pt_nM ($n = 6, 9$; M = Pt, Ru, Sn), the reaction pathways for each alloy catalyst are compared with respect to thermodynamics and kinetics, and the overall reaction rate.

3.2 Theoretical Methodology and Bimetallic Models

Both metal slab and cluster are widely used to model nanoparticles of metal catalysts. The plane-wave-based periodic density functional theory (DFT) was mainly applied to the former, and basis set expansion DFT to the latter as used in the current chapter. For all the involved metal atoms LANL2DZ type of effective core potential (ECP) and the corresponding double-ζ basis set were employed, while the basis sets for C, H, and O are 6-311++G(d,p). In the LANL2DZ ECP scheme valence electron shells for the ground state of Pt, Sn, Ru, Rh, Pd, Cu, and Re are $5s^25p^65d^96s^1$, $5s^25p^2$, $4s^24p^64d^65s^2$, $4s^24p^64d^75s^2$, $4s^24p^64d^85s^2$, $3s^23p^63d^{10}4s^1$, and $5s^25p^65d^56s^2$, respectively.

To discuss the decomposition of H_2O over nanoscale particle catalysts that usually have a well-defined fcc surface, the one layer extend cluster model Pt_6M as well as the two layer model $(Pt_6M)Pt_3$ have also been applied, which were shown in Fig. 3.1. Constraint optimization was employed for H_2O adsorption on the surface models with the fixed cluster. The frequency analysis usually presents a few small imaginary frequencies due to the fixed cluster structure, while for the relevant TS structures of adsorbed water dissociation there is a unique significant imaginary frequency that can be assigned to OH breaking. The transition states (TS) for

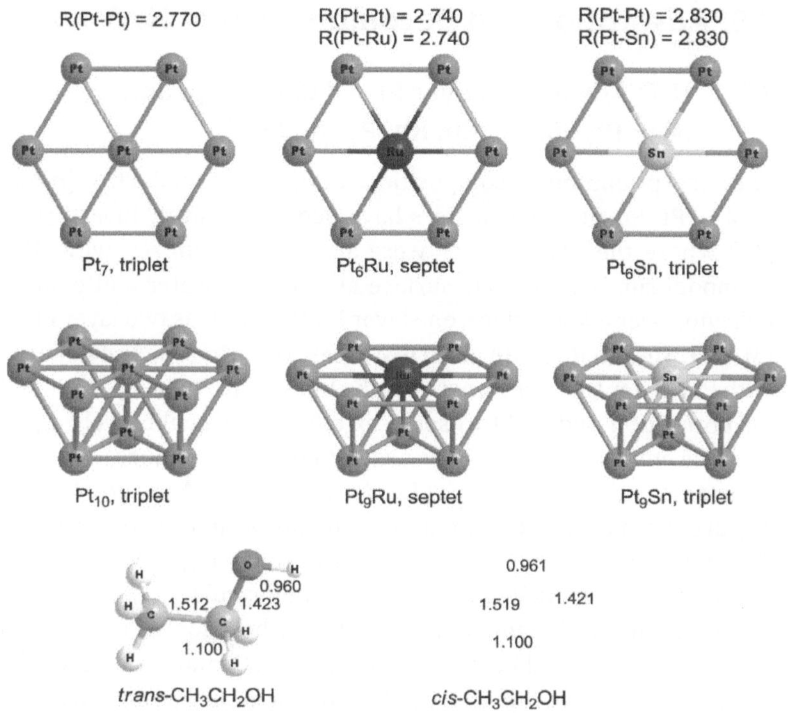

Figure 3.1 Structures of ethanol and Pt_nM ($n = 6$ and 9; M = Pt, Ru, Sn) models with some bond length parameters in Å. Reprinted with permission from Xu and Wang [18]. Copyright © 2013 American Chemical Society.

dissociations of the adsorbed water and ethanol were directly searched for by either the transit-guided quasi-Newton (STQN) method or the eigenvector following with the Berny algorithm. A frequency analysis was done at the same level as optimization for all stationary points to characterize the stationary points and make the zero-point energy (ZPE) corrections.

Water/ethanol adsorption energies (E_{ads}) are defined as $E_{ads} = E$(H_2O/ethanol-Pt_nM in ground state)-[E(H_2O/ethanol)+E(Pt_nM in ground state)], the difference between energy of the adsorption complex and the energy summation of the cluster at ground state and water or ethanol. The higher negative E_{ads} indicates the stronger adsorption between water/ethanol and the cluster.

3.3 Adsorption and Decomposition of H_2O

3.3.1 H_2O Adsorption over Pt_n M Clusters (6 and 9; M = Pt, Sn, Ru, Cu, Rh, Pd, and Re)

The adsorptions and decompositions of H_2O over small Pt_nM ($n = 2$, 3; M = Pt, Ru, Sn, etc.) clusters have been extensively investigated [13] where the Pt_nM cluster were also fully optimized with DFT. To model sub-nanoparticle surface and discuss water and ethanol behavior over alloy surface, one layer Pt_6M as well as two layer Pt_9M (M = Pt, Sn, and Ru) with given distances were chosen. As compared with Pt cluster the bond lengths of the bimetallic Pt_nSn are extended for both Pt-Sn and Pt-Pt to some extent, which qualitatively agree with the empirical equation for nearest neighbor distances R of PtM bimetallic cluster [19], $R = (R_M/R_{Pt})^{1/3} \times 2.77$ Å, where R_M and R_{Pt} are atomic radii for the alloyed metal M and Pt, respectively. Therefore, the shortest distance in the following PtM bimetallic clusters will be estimated by the empirical formula.

Pt_7 is one layer model consisting of a hexagon of 6 Pt atoms surrounding a central Pt atom; while in the Pt_6M bimetallic alloy the central atom will be replaced by alloyed atoms such as Sn and Ru. To eliminate the edge effect, a two layer cluster, $(Pt_6M)Pt_3$, with 7 atoms (Pt_6M) on the surface layer and 3 Pt atoms on the second layer, has also been used to model Pt-M alloy surface, as employed by Ishikawa et al. [19].

Water adsorptions on both central and peripheral Pt atoms of Pt_7 are explored. The adsorption on the peripheral Pt is slightly stronger than on the central Pt (E_{ads}: −0.46 vs. −0.43 eV), and the weaker adsorption on the central Pt accompanies a slight larger Pt-O distance by 0.01 Å (Pt-O: 2.347 vs. 2.331 Å). Both adsorptions take atop configuration (Figs. 3.2a and 3.2b) with O-Pt-Pt angle in a range of 85–95° and water almost parallel to Pt(111)surface. The binding energy of H_2O on the central Pt is well within the literature reported results for Pt(111) surface with a variety of theoretical methods and models, −0.27 eV [20], −0.29 eV [21], −0.33 eV [22], −0.42 eV [19], and −0.56 eV [23].

The nearest neighbor distance in the one layer Pt_6Sn adopt the experimental value, 2.83 Å [19]. Starting with atop (Sn), bridge

(Sn-Pt), as well as hollow (Sn-Pt-Pt) adsorptions via the O atom, only atop adsorption was located for the Pt_6Sn surface. The adsorption in the central Sn becomes remarkably strong with adsorption energy of -1.30 eV and Sn-O distance of 2.305 Å. To investigate the edge effect of Sn on water adsorption, one peripheral Pt atom in Pt_7 was also replaced by Sn. Water adsorption on the peripheral Sn ($E_{ads} = -0.61$ eV; Sn-O: 2.417 Å; Sn-H: 2.73 and 2.79 Å) is turned out to be much weaker than that on the central Sn atom. As shown in Figs. 3.2c and 3.2d, the primary adsorptions on central and outside Sn of the Pt_6Sn surface are atop, where oxygen sits right above Sn with O-Sn-Pt in the range of 82–97° and water is roughly parallel to the Pt_6Sn (111) surface.

The nearest neighbor distance in the one layer Pt_6Ru was 2.74 Å, estimated by the above mentioned empirical formula. Adsorption on the central Ru (2e) is slightly weaker than on Sn (E_{ads}: -1.25 vs. -1.30 eV) yet much stronger than that on the central Pt of Pt_7 (E_{ads}: -1.25 vs. -0.43 eV).

To address the three-dimensional structure effect, the second layer with three Pt atoms was added in (111) pattern. For the two-layer cluster Pt_{10}, the adsorption strength on the peripheral Pt atom is similar to that of one-layer Pt_7 (E_{ads}: -0.48 vs. -0.46 eV); however, in spite of similar atop adsorption pattern and slightly longer Pt-O distance (2.367 vs. 2.347 Å) the adsorption on the central Pt of Pt_{10} is much weaker than that of Pt_7 (E_{ad}: -0.26 vs. -0.43 eV). The predicted binding energy of H_2O on the central Pt atom of Pt_{10} in this work, -0.26 eV, is rather close to the plane-wave periodic DFT calculations of monomer water on Pt(111) surface, -0.29eV, [21] and -0.33 eV [22]. Although the same cluster model, two-layer Pt_{10}, has been employed by Ishikawa et al. [19] and the present investigation, the present adsorption is weaker than that predicted with the ADF method there (-0.26 vs. -0.46eV).

As the surface of Pt_{10} was alloyed with Sn in the center, similar to the cases of Pt_6Sn and Pt_7 the adsorption enhancement still remains due to alloying Sn. The water adsorption energy on central Sn of Pt_9Sn is much stronger than that on central Pt of Pt_{10} (E_{ads}: -0.80 vs. -0.26 eV). Compared with the adsorption on Pt_6Sn, the introduction of the second year in Pt_9Sn weakens water atop adsorption on the Sn site (E_{ads}: -0.80 vs. -1.30 eV). The result is significantly stronger

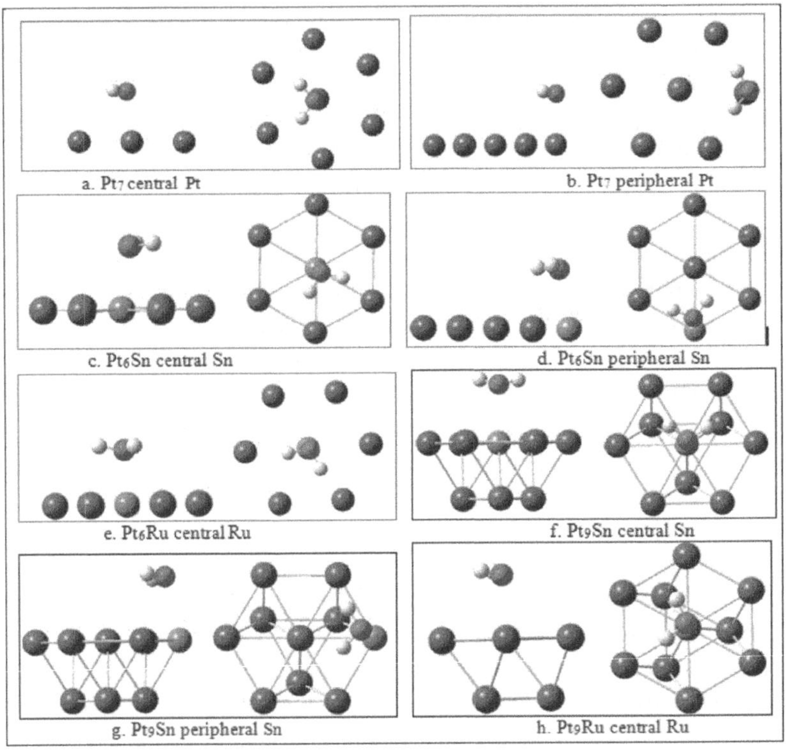

Figure 3.2 Water atop adsorption on M site of Pt_6M and Pt_9Sn. In each panel the first one is side view and the second top view. Reprinted with permission from Wang et al. [13]. Copyright © 2010 American Chemical Society.

than the unique available reference to our knowledge for H_2O adsorption on Sn site (−0.80 vs. −0.39 eV [19]). Atop adsorption on the peripheral Sn (Fig. 3.2g) was also located ($E_{ads} = -0.32$ eV, Sn-O = 2.579 Å) and is significant weaker than the atop adsorption on the central Sn, which has the same trend as the one-layer Sn alloying cluster Pt_6Sn, yet opposite to the central and peripheral Pt atoms of Pt_{10}.

Likewise, water adsorption on the central Ru site of the Pt_9Ru surface also presents atop configuration with adsorption energy of −0.39 eV and Ru-O distance of 2.425 Å. It is interesting to note that water adsorption strength has the same trend as the above one-layer models as compared with Pt and Sn alloying surfaces, i.e., the adsorption on Ru atop site is also in-between those on Sn of Pt_9Sn ($E_{ads} = -0.80$ eV, Sn-O = 2.085 Å) and on Pt of Pt_{10} ($E_{ads} = -0.26$ eV, Pt-O = 2.367 Å).

The above two-layer model $(Pt_6M)Pt_3$ was also extended to alloying Cu, Rh, Pd, and Re, all of which have lower first IP than Pt. Therefore, electron transfer should occur from the central atoms to the surrounding Pt atoms. The total spin (S) of the ground state is 3/2, 1/2, 1, and 5/2 for bimetallic PtCu, PtRh, PtPd, and PtRe, respectively. In all of the cases, H_2O preferentially binds to the central atom M in atop configuration with the dipole moment of water almost parallel to the cluster surface. The binding energy of H_2O on Re site of Pt_9Re (Fig. 3.2h) is stronger than that on Sn site of Pt_9Sn (E_{ads}, -1.13 vs. -0.80 eV), while all of others are weaker. In addition, E_{ads} on Re in this work is the same as the reported value by Ishikawa et al. [19]. The current adsorption energy on the Cu site of Pt_9Cu (E_{ads}, -0.52 eV) is close to Ishikawa's result (E_{ads}, -0.66 eV), but that on Rh is much weaker than that predicted by Ishikawa et al. (E_{ad}, -0.45 vs. -0.81 eV). The binding energy on Pd site (-0.37 eV) is comparable to that on Pd(111) (-0.33 eV) [22]. Water adsorption strength on the central M of Pt_9M decreases in the sequence, Re (-1.25 eV) >Sn (-0.80 eV) >Cu (-0.52 eV) >Rh (-0.45) >Ru (-0.39) >Pd (-0.37 eV) >Pt (-0.26 eV). The adsorption strength seems to have a weak correlation with the valence *d* orbital vacancy of ground state M. For instance, the high vacant *d* shell of Re and Sn ($5d^5 6s^2$ and $5s^2 5p^2 5d^0$) accompany strong water adsorption; on the other hand, water adsorption on the low vacancy of Pd and Pt ($4d^{10}$ and $5d^9 6s^1$) is weak.

3.3.2 Decomposition of Adsorbed Water

For the adsorbed water on the central metal of Pt_6M (111) surfaces (M = Pt, Sn, and Ru), dissociation kinetics and thermodynamics agree well with the trends for the optimized Pt_2M [17]. According to Table 3.1, dissociation energy barrier for the adsorbed water over Sn is slightly lower than over Ru but remarkably lower than over the central Pt (E_a: 0.60 vs. 0.67 and 1.14 eV). In spite of dramatically smaller than in the gas phase (5.26 eV), for all of three

Table 3.1 Water adsorption and decomposition over **M** site of Pt_nM ($n = 6$ and 9; M = Pt, Sn, Ru, Rh, Pd, Re, and Cu)[a]

					Dissociation $H_2O^* \rightarrow H^* + HO^*$				
	Adsorption complex[b]				Transition state			Energetic data	
	M	R (O–M)	E_{ads}	q	ν	R (O–M)	R (O–H)	E_a	E_{diss}
H_2O									5.26
Pt_7*_a	5	2.347	−0.43 (−0.34)	0.44	844	2.044	1.512	1.14 (0.98)	0.90
Pt_7*_b	3	2.332	−0.46 (−0.36)	0.33					
Pt_6Sn*_a	3	2.305	−1.30 (−1.19)	0.37	425	2.083	1.529	0.60 (0.45)	0.50 (0.37)
Pt_6Sn*_b	3	2.417	−0.61 (−0.53)						
Pt_6Ru*_a	7	2.116	−1.25 (−1.12)	0.51	880	1.973	1.431	0.67 (0.50)	0.32 (0.18)
Pt_{10}*_a	3	2.367	−0.26 (−0.14)	0.32	226	2.062	1.726	1.07 (0.93)	0.67
Pt_{10}*_b	3	2.255	−0.48 (−0.35)	0.16	1085	2.095	1.480	1.19 (1.03)	
Pt_9Sn*_a	3	2.402	−0.80 (−0.72)	0.33	412	2.085	1.539	0.78 (0.64)	0.41 (0.28)
Pt_9Sn*_b	5	2.579	−0.32						
Pt_9Ru*	7	2.425	−0.39 (−0.28)	0.39	522	2.048	1.554	0.96 (0.81)	0.61 (0.50)
Pt_9Rh*	2	2.200	−0.45	0.53	346	2.017	1.660	1.18	0.87
Pt_9Pd*	3	2.364	−0.37	0.38	308	2.027	1.848	1.36	1.20
Pt_9Re*	4	2.192	−1.25	0.42	1127	2.042	1.378	0.71	−0.43
Pt_9Cu*	4	2.414	−0.52	0.47	616	1.949	2.146	1.81 (1.60)	1.60

[a] E_{ads} (eV): adsorption energy on M site; M: 2S+1; R(O–M/Å): distance between H_2O oxygen and M; q(e): charge carried by adsorbed water; ν (cm^{-1}) and E_a (eV): imaginary frequency and activation energy for the transition state of adsorbed water decomposition; E_{diss} (eV): dissociation energy for the adsorbed water. One asterisk refers to an adsorption site.
[b] Pt_nM*_a and Pt_nM*_b stand for adsorptions on the central and peripheral M atoms, respectively.
Adapted with permission from Wang et al. [13]. Copyright © 2010 American Chemical Society.

cases dissociation energies of adsorbed water on Pt_6M (111) are still positive (0.50, 0.32, and 0.90 eV). Dissociated OH and H groups are atop adsorbed on M and Pt with O–M distances 0.1 Å less than corresponding transition states. In the case of Pt_6M, all of the TS were turned out to have a major imaginary frequency characterizing OH breaking of the adsorbed water.

For the extended two-layer (111) surfaces Pt_9M (M = Pt, Sn, Ru, Cu, Rh, Pd, and Re), the adsorbed water on the central M of the top layer exhibits atop configuration. The transition states of adsorbed water on Pt and Ru are later as compared those of Pt_6M, which was evidenced by longer O–H distances. Consequently, the corresponding imaginary frequencies of the TS are smaller. For Sn, the characteristics of the TS are very comparable to one-layer Pt_6Sn. E_a for the central Pt atom of Pt_{10} is slightly lower than that of Pt_7 (1.07 vs. 1.14 eV), while the barriers for Sn and Ru are increased to different magnitude due to the second layer effect (0.78 vs. 0.60 eV for Sn; 0.96 vs. 0.67 eV for Ru). According to Table 3.1 and Fig. 3.3, the energy barriers for Pt_9M (M = Pt, Sn, and Ru) have the same trend as the above one-layer model Pt_6Sn and small optimized clusters Pt_nM. Once again the adsorbed water on Sn shows the lowest dissociation energy barrier, followed by Ru and Pt (E_a: 0.79 vs. 0.96 and 1.07 eV). It is interesting to note that the dissociation energy on Sn site is also the lowest (E_{diss}: 0.41 vs. 0.61 and 0.67 eV). This implies that from both kinetic and thermodynamic viewpoints Sn is the most active to water decomposition among the pure Pt and two PtM (M = Sn and Ru) alloys, which well supports the assumption of the bifunctional mechanism that Sn site accelerates the dissociation of H_2O.

We acknowledge that the absolute energetic data for the individual Pt_nM cluster may not be accurate, but the comparative study on the base of the bimetallic cluster still well fulfils the present objective. No matter how large the theoretical model is, the critical issue is whether the applied model provides consistent results and could provide new insight into the objective. As shown by Table 3.2, three distinct models ranging from 3 to 10 atoms and one to two layers qualitatively show the same trend that Sn has the lowest E_a for water decomposition, closely followed by Ru and Pt with the greatest E_a. Quantitatively, the relative ΔE_a values of PtSn, PtRu, and Pt are also comparable for the three models. The consistent trend

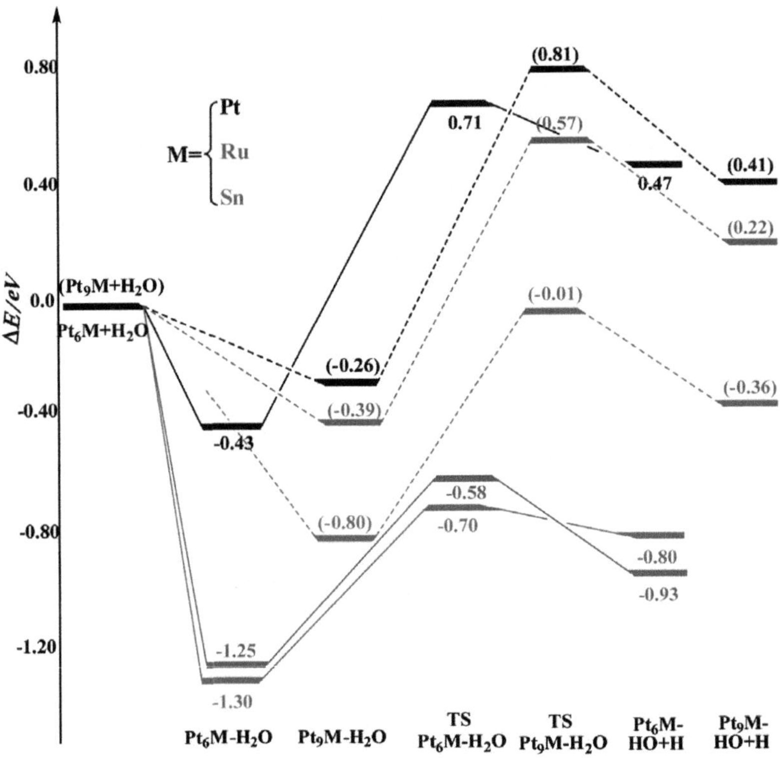

Figure 3.3 Potential energy profile for water adsorption and dissociation on the M site of Pt_nM ($n = 6$ and 9; M = Pt, Ru, and Sn). The solid and dotted lines link the species from water adsorption and dissociation on Pt_6M and Pt_9M, respectively. Reprinted with permission from Wang et al. [13]. Copyright © 2010 American Chemical Society.

provides theoretical evidence for the fact that Sn is more active to water decomposition than Ru and Pt.

The present E_a for water decomposition on Pt_{10} (1.07 eV) is slightly higher than the reported values with a variety of theoretical methods and models, such as 1.02 eV with ADF code and two-layer Pt_{10} cluster [19], 0.68–0.88 eV on Pt(111) with plane-wave periodic DFT (RPBE or PW91) and ultrasoft pseudopotential (USPP) implemented on VASP, DACAPO or CASTEP [22, 24, 25]. The E_a on Sn (0.79 eV) is lower by 0.13 eV than that (0.92 eV) calculated with the two-layer Pt_9Sn model and ADF code [19], while others

Table 3.2 Variation of activation energy of water decomposition on Pt_nM

	M	E_a (eV)	ΔE_a (eV)
Pt_2M	Sn	0.61	0.0
	Ru	0.63	0.03
	Pt	1.14	0.53
Pt_6M	Sn	0.60	0.0
	Ru	0.67	0.07
	Pt	1.14	0.54
Pt_9M	Sn	0.78	0.0
	Ru	0.96	0.18
	Pt	1.07	0.29

are systematically higher by 0.05~0.54 eV than those from Ishikawa et al. (E_a: 0.71 vs. 0.66 eV for Re; 0.96 vs. 0.84 eV for Ru; 1.81 vs. 1.38 eV for Cu; 1.18 vs. 1.00 eV for Rh). The E_a on Pt_9Pd (1.36 eV) is rather comparable to literature data on Pd(111), e.g., 1.12 eV [26], 1.09 [27], and 1.05 eV [22]. Water dissociation kinetic activity increases in the sequence of Cu <Pd <Rh <Pt <Ru <Sn <Re. In spite of the difference, the sequence is somewhat relevant to water adsorption strength order, Pt<Pd<Ru<Rh<Cu<Sn<Re. Taking Pt as a reference, the plot of relative E_a against relative E_{diss} for water dissociation over the investigated transition metals M (Pd, Ru, Rh, Cu, Sn, and Re) is illustrated in Fig. 3.4. According to the plot, the bimetallic alloys PtM (M = Re, Sn, and Ru) are more active to water dissociation both kinetically and thermodynamically, while the bimetals alloying with Cu, Pd, and Rh are less active than Pt from both aspects.

Water adsorption on the Sn site of Pt_nSn ($n = 2$ and 3) is weaker than those on the Pt site of Pt_n ($n = 3$ and 4) and the Ru site of Pt_nRu ($n = 2$ and 3), while water adsorptions on the central Sn atom of Pt_6Sn and Pt_9Sn are enhanced so significantly that they are stronger than those on the central Pt and Ru atoms of Pt_nM ($n = 6$ and 9; M = Pt and Ru). For all of the three cluster models, energy barrier (E_a) for the dissociation of adsorbed water over Sn is lower than

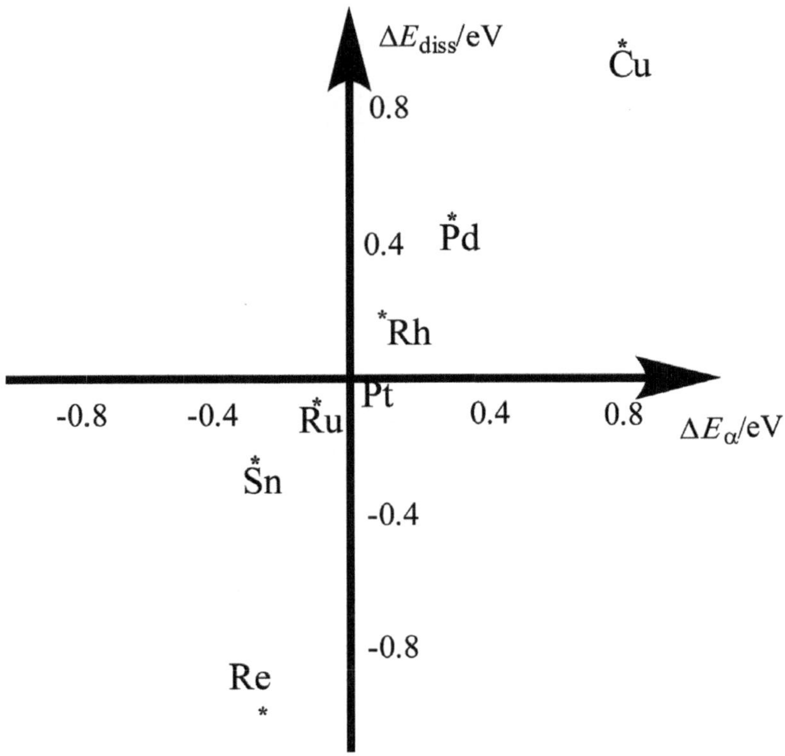

Figure 3.4 The plot of the relative dissociation energy against the relative energy barrier for water dissociation over the M site of Pt_9M. Reprinted with permission from Wang et al. [13] Copyright © 2010 American Chemical Society.

over Ru and Pt atoms (e.g., E_a: 0.78 vs. 0.96 and 1.07 eV for Pt_9M). It is interesting to note that the dissociation energy on Sn site is also the lowest (E_{diss}: 0.44 vs. 0.61 and 0.67eV). The results show that from both kinetic and thermodynamic viewpoints Sn is more active to water decomposition than pure Pt and the other investigated PtM alloys except for PtRe, and the kinetic activity for water dissociation increases in the sequence of Cu $<$Pd $<$Rh $<$Pt $<$Ru $<$Sn $<$Re, which well supports the assumption of the bifunctional mechanism that Sn site accelerates the dissociation of H_2O. However, although the energy barrier for water decomposition on the Re site of the PtRe is only slightly lower than that on Sn site, the adsorption energy and dissociation energy on the Re site are considerably more favorable than those on the Sn site of PtSn. Thus, the reason why the PtRe is not as active to the EER as PtSn remains unclear.

3.4 Adsorption and Decomposition of Ethanol

3.4.1 CH_3CH_2OH Adsorption and Decomposition over Pt_6M Clusters (M = Pt, Ru, and Sn)

To further understand the bifunctional mechanism, the adsorptions and dehydration of ethanol catalyzed by the bimetallic were also extensively investigated. *Trans*-C_2H_5OH has slightly lower energy by 0.0003 eV than the *cis*-C_2H_5OH (Fig. 3.1), and they can isomerize readily each other because of a low energy barrier 0.046 eV at the B3PW91/6-311++G(d, p) level. The optimized geometries of ethanol adsorbed on Pt_7, Pt_6Ru, and Pt_6Sn are shown in Figs. 3.5, 3.6, and 3.7, respectively, while their relative energy profiles are illustrated in Fig. 3.8.

Three kinds of adsorption of ethanol on Pt_7 are considered through both hydroxyl and methylene groups. On the first one (Pt_7-ethanol-a), the CH_2 of *trans*-CH_3CH_2OH is directly adsorbed via the alpha hydrogen (H_α) on the central Pt, accompanied by a weak adsorption of hydroxyl hydrogen on one edge Pt. The distances of H_α-Pt and hydroxyl H-Pt are 1.853 and 2.426 Å, respectively. The bond lengths of C_α-H_α and H-O are elongated by 0.093 and 0.021 Å, respectively, compared to those of *trans*-CH_3CH_2OH. The adsorption energy is predicted to be −0.12 eV, as shown in Fig. 3.8. The adsorbed complex (Pt_7-ethanol-a) decomposes to the adsorbed product (Pt_7-ethanol-P-a) via the transition state (Pt_7-ethanol-TS-a, imaginary frequency 1914 cm^{-1}). On the TS the C–H_α significantly and O–H slightly elongate to 1.528 and 1.026 Å, while the C–O shortens to 1.299 Å; the H_α and hydroxyl H get closer to Pt atoms by approximately 0.3 Å. It is interesting to note that the major eigenvectors correspond to the stretching of both C–H_α and O–H bonds, which indicates the simultaneous cleavage of the two bonds, and a visualization clearly shows that the TS (Pt_7-ethanol-TS-a) connects the adsorbed complex Pt_7-ethanol-a and the dehydrogenation product Pt_7-ethanol-P-a. The *concerted* dehydrogenation pathway agrees well with Wang's report [16]; however, it was attributed to the adsorption complex via the hydroxyl instead of C–H_α adsorption. On the product Pt_7-ethanol-P-a, the H_α atom migrates from the central Pt to an edge Pt forming a H-Pt bond of 1.507 Å, and the broken O–H forms a hydrogen bond with H-O distance of 1.796 Å and H-Pt distance of 1.535 Å. Acetaldehyde, CH_3CHO, is generated through the concerted path. According to Fig. 3.8, the energy barrier E_a and dissociation energy E_{diss} with respect to the adsorbed ethanol (Pt_7-ethanol-a) are 0.39 and -0.94 eV, respectively.

The second adsorption occurs with *trans*-CH_3CH_2OH on Pt_7 to form complex (Pt_7-ethanol-b; Fig. 3.5) via the hydroxyl group. The O binds to the central Pt through lone electron pairs and its H atom connects to one edge Pt. On Pt_7-ethanol-b the O-Pt and H-Pt distances are 2.361 and 2.733 Å, respectively. The adsorption of Pt_7-ethanol-b ($E_{ads} = -0.29$ eV) is much stronger than that of Pt_7-ethanol-a ($E_{ads} = -0.12$ eV), which may be due to electrostatic attraction between the negatively charged O atom (-0.280e) and the positive central Pt atom ($+0.387$e). In the dehydrogenation process from Pt_7-ethanol-b passing over the transition state (Pt_7-ethanol-TS-b, imaginary frequency 714 cm^{-1}) to the product (Pt_7-ethanol-P-b), the O–H bond elongates from 0.970 Å to 1.524 Å and finally breaks; meanwhile, the H-Pt distance decrease from 2.733 Å to 1.602 Å on the TS and to 1.505 Å on the product, and the O-Pt distance changes from 2.361 Å to 2.084 Å and finally to 1.981 Å. The dehydrogenation process has much higher energy barrier than the previous concerted pathway initiated by C–H_α adsorption (E_a: 1.09 vs. 0.39eV), and thermodynamically the dissociation energy is also less favorable (E_{diss}: 0.77 vs. -0.94 eV).

The adsorption and subsequent dissociation of *cis*-CH_3CH_2OH on Pt_7 via the hydroxyl group are also investigated. As shown in Fig. 3.5, the adsorbed complex (Pt_7-ethanol-c) adopts a titled atop adsorption geometry via oxygen with OH almost parallel to the surface, which is similar to the widely reported conformation for methanol as well as ethanol adsorptions over transition metal clusters or surfaces [14, 28–30]. The distances of O-Pt and H-Pt at the adsorption complex (Pt_7-ethanol-c) are 2.329 and

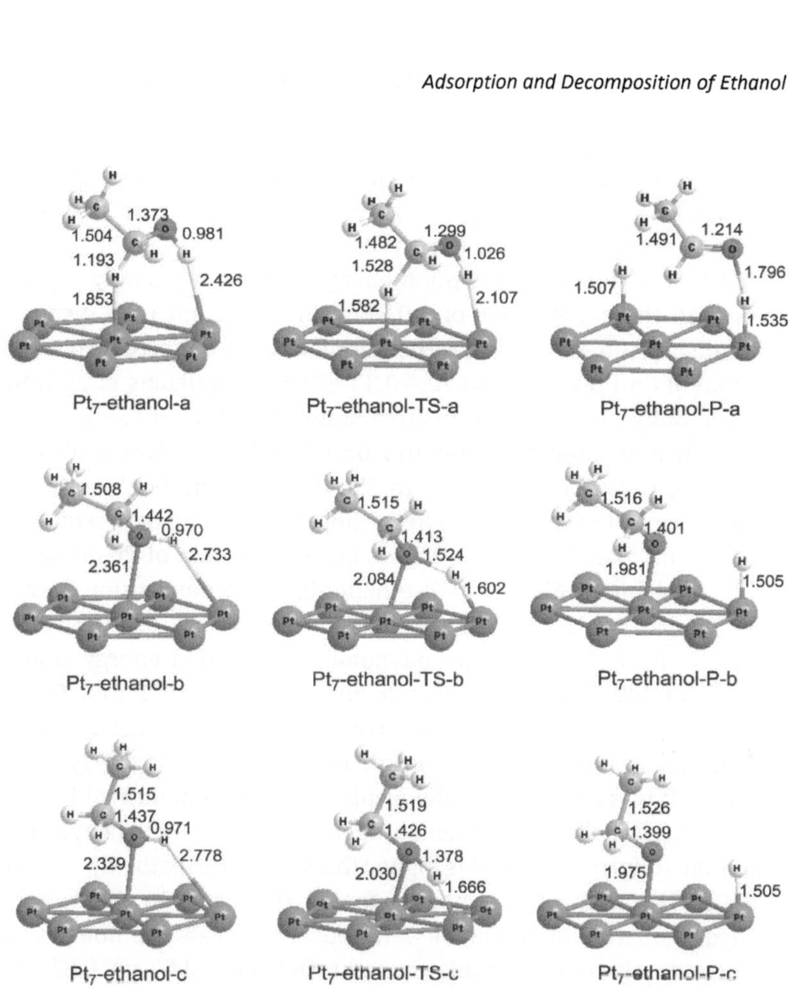

Figure 3.5 Structures of stationary points of Pt_7+ CH_3CH_2OH with some critical bond length parameters in Å. Reprinted with permission from Xu and Wang [18]. Copyright © 2013 American Chemical Society.

2.778 Å, respectively. The former is slightly shorter than that of Pt_7-ethanol-b and the latter is slightly longer than that of Pt_7-ethanol-b. Although the *trans*-CH_3CH_2OH has slightly lower energy than the *cis*- CH_3CH_2OH, the adsorbed *cis*-CH_3CH_2OH (Pt_7-ethanol-c) is more stable than the corresponding complex from the former (Pt_7-ethanol-b). Thus, among the three adsorption conformations over Pt surface, the Pt_7-ethanol-c is the most stable one with a binding energy of -0.32 eV, stronger than Pt-ethanol-b ($E_{ads} = -0.29$ eV) and Pt-ethanol-a ($E_{ads} = -0.12$ eV) by 0.04 and 0.20 eV, respectively.

Alcala et al. found that *cis*-ethanol binds through oxygen at a top site on Pt(111) [14], in good agreement with the current result. Their calculated O-Pt bond length (2.59 Å) is longer than the present one (2.33 Å), but the adsorption energy (−0.29 eV) is only slightly less than the current one of −0.32 eV. The present O-Pt distance and adsorption energy are rather similar to the reported values of methanol on Pt(111) (2.43 Å, −0.33 eV) by Mavrikakis et al. [28]. On the dehydrogenation transition state (Pt_7-ethanol-TS-c), the O-Pt and H-Pt distances shorten to 2.030 Å and 1.666 Å, respectively, while the O–H bond elongates to 1.378 Å, also shorter by 0.146 Å than that of Pt_7-ethanol-TS-b.Accordingly, the imaginary frequency ($i1157$ cm^{-1}) of Pt_7-ethanol-TS-c is larger than that of Pt_7-ethanol-TS-b ($i714$ cm^{-1}). Similar to Pt-ethanol-b, its decomposition energy barrier (E_a = 1.07 eV) is much higher than that of Pt-ethanol-a (E_a = 0.39 eV), and thermodynamic dissociation energy is also less favorable than that of Pt-ethanol-a (E_{diss} = 0.71 eV). Above all, over Pt_7 surface the most favorable dehydrogenation path is the C–H_α cleavage in terms of kinetic as well as thermodynamic aspects. It was also reported by Alcala et al. [31] that on Pt(111) the energy barrier for the first dehydrogenation of ethanol via hydroxyl cleavage to form CH_3CH_2O is higher than the C–H_α pathway to form CH_3CHOH (relative to gas phase ethanol, 0.53 vs. 0.46 eV). For the dehydrogenation of methanol, C–H cleavage was also demonstrated to be more kinetic favorable than the O–H (E_a: 0.67 vs. 0.81 eV) [28]. Our results basically supports the speculation that the first dehydrogenation of ethanol may proceed through the cleavage of C–H_α [32], rather than OH [33].

On Pt_6Ru ruthenium atom replaces the central Pt of Pt_7. Because of more unoccupied *d* orbitals on Ru, the bimetallic cluster Pt_6Ru might have different ligand effect on ethanol than Pt_7. The adsorption energy for Pt_6Ru-CH_3CH_2OH through the α-hydrogen is predicted to be −0.61 eV, more exothermic than Pt_7-CH_3CH_2OH by 0.49 eV. As shown in Fig. 3.6, the distances of H_α-Ru and C_α-Ru in the complex, Pt_6Ru-ethanol-a, are 1.684 Å and 2.602 Å, respectively, much shorter than those on Pt_7-ethanol-a complex. This strong adsorption considerably weakens the C–H_α bond to 1.234 Å. The energy of the transition state (Pt_6Ru-ethanol-TS-a from Pt_6Ru-ethanol-a to the product complex Pt_6Ru-ethanol-P-a) is

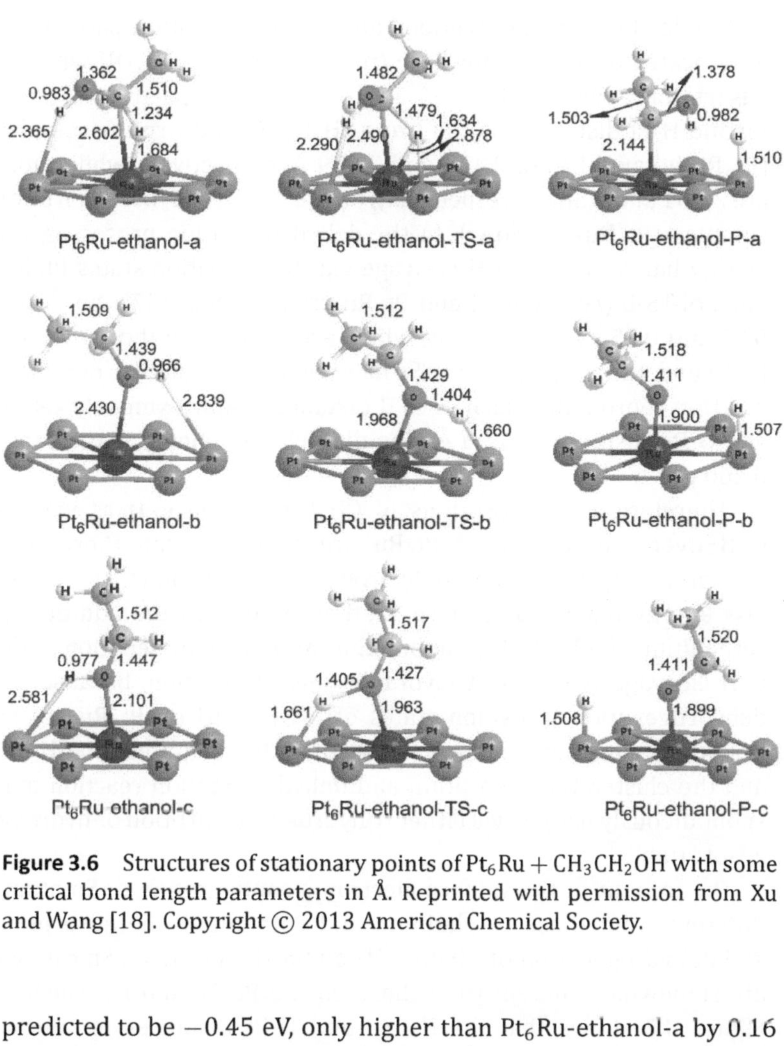

Figure 3.6 Structures of stationary points of $Pt_6Ru + CH_3CH_2OH$ with some critical bond length parameters in Å. Reprinted with permission from Xu and Wang [18]. Copyright © 2013 American Chemical Society.

predicted to be −0.45 eV, only higher than Pt_6Ru-ethanol-a by 0.16 eV with an imaginary frequency 600 cm^{-1}. The C–H_α bond on Pt_6Ru-ethanol-TS-a is extended to 1.479 Å, and then is completely broken and α-H shifts to the neighbor Pt with C-Ru and H_α-Pt bond lengths of 2.144 and 1.510 Å, respectively. The C–H_α cleavage over Ru has a rather low energy barrier of 0.16 eV. However, Pt_6Ru-ethanol-P-a shows that only the α-H atom is removed from CH_3CH_2OH in this dehydrogenation reaction, rather than a concert C–H_α and O–H cleavage as in the case of Pt_7-ethanol.

Similar hydroxyl adsorptions and dehydrogenation paths were also located over Pt_6Ru for both *trans*- and *cis*-CH_3CH_2OH with the adsorption energies of −1.06 and −1.08 eV, respectively. The O-Ru and H-Pt distances on the *trans*-CH_3CH_2OH adsorption complex (Pt_6Ru-ethanol-b) are 2.430 and 2.839 Å, respectively, while those are 2.101 and 2.581 Å, respectively, on the *cis*-CH_3CH_2OH adsorption complex (Pt_6Ru-ethanol-c). In the dehydrogenation processes, the energy barriers of the OH cleavage via the transition states Pt_6Ru-ethanol-TS-b (i900 cm^{-1}) and Pt_6Ru-ethanol-TS-c (i875 cm^{-1}) are 0.57 and 0.55 eV, respectively, which are lower than those over Pt_7 (1.09 and 1.07 eV). The two TSs have very rather similar dominant reaction coordinates, such as O–H distance of approximately 1.40 Å and O-Ru distance of 1.96 Å. The dissociation energy is the same, 0.260 eV.

Therefore, the adsorptions of CH_3CH_2OH via C–H as well as O–H over the Ru site of Pt_6Ru are stronger than those over Pt_7, and the subsequent dehydrogenation reactions also show less energy barrier and more exothermic due to the addition of ruthenium. Similar to Pt_7, among the three dehydrogenation paths C–H cleavage is the most favorable one. In addition, because the dehydrogenation transition states of CH_3CH_2OH on Pt_6Ru are all predicted to have much lower energy than the separated CH_3CH_2OH and the cluster, the adsorption and dehydrogenation reaction may spontaneously happen via either α-hydrogen adsorption or hydroxyl adsorption.

On account of the valence electronic configuration of Sn atoms different from those of platinum and ruthenium, the adsorption and dehydrogenation of CH_3CH_2OH on Sn site of the Pt_6Sn catalyst are somewhat different from those on the Pt_6Ru and Pt_7 clusters. Figure 3.7 shows that the H_α-Sn and H-Pt distances for the α-hydrogen absorption complex (Pt_6Sn-ethanol-a) are 2.470 and 2.361 Å, respectively; and that the C_α atom is far from the Sn atom. The adsorption energy is −0.74 eV, stronger than Pt_7-ethanol-a and Pt_6Ru-ethanol-a as shown in Fig. 3.8. Similar to the corresponding adsorption on Pt_7, the complex Pt_6Sn-ethanol-a also leads to a concerted O–H and C_α-H_α cleavage. However, its dehydrogenation transition state (Pt_6Sn-ethanol-TS-a) is of a very high energy, 0.31 eV relative to $Pt_6Sn + CH_3CH_2OH$, with a rather high energy barrier

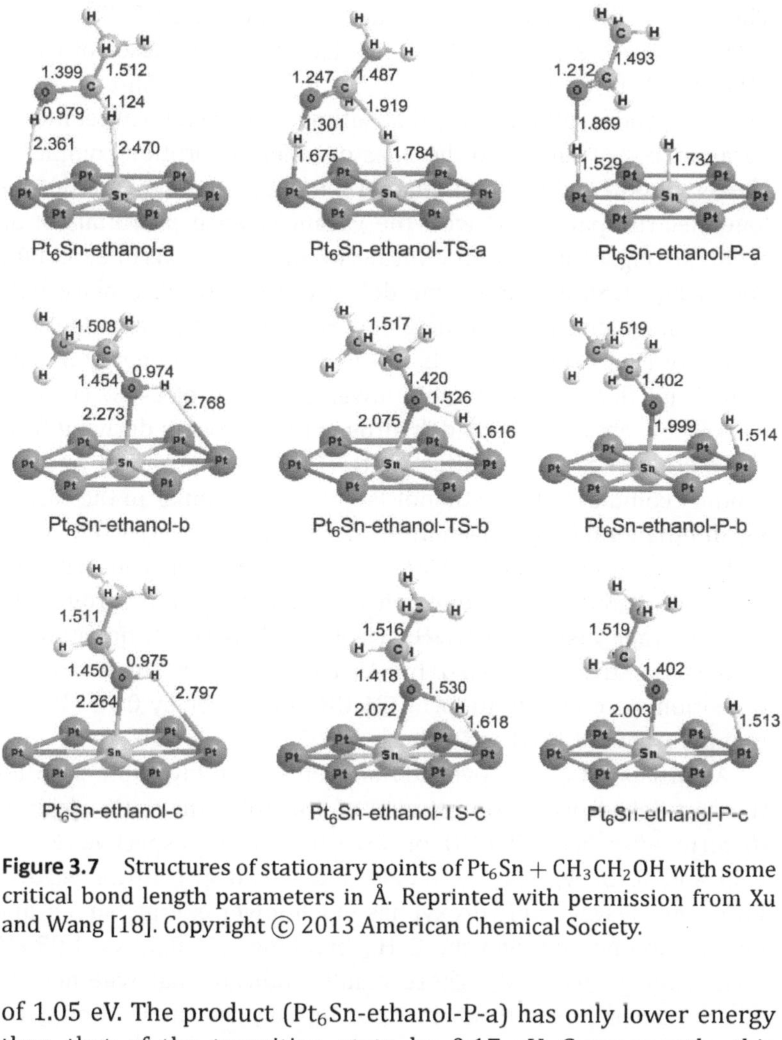

Figure 3.7 Structures of stationary points of $Pt_6Sn + CH_3CH_2OH$ with some critical bond length parameters in Å. Reprinted with permission from Xu and Wang [18]. Copyright © 2013 American Chemical Society.

of 1.05 eV. The product (Pt_6Sn-ethanol-P-a) has only lower energy than that of the transition state by 0.17 eV. Consequently, this reaction channel is an endothermic and may be a non-spontaneous process.

The adsorption and dehydrogenation of CH_3CH_2OH through the hydroxyl adsorption type is more feasible. The adsorption energies of *trans*- and *cis*-CH_3CH_2OH on Pt_6Sn through the hydroxyl are predicted to be −1.36 and −1.40 eV, respectively, which are the most strong adsorptions of the investigated cases in Fig. 3.7.

The O-Sn (H-Pt) distances are approximately 2.27(2.77) Å and 2.26 (2.80) Å, respectively, on the adsorption complexes of *trans*-CH_3CH_2OH (Pt_6Sn-ethanol-b) and *cis*-CH_3CH_2OH (Pt_6Sn-ethanol-c). The strong adsorptions imply that, although these critical geometric parameters are similar to the ones of other adsorption complexes (Pt_7-ethanol-b and c, Pt_6Ru-ethanol-b and c), the interaction of the lone electron pairs of O with the vacant valence *p* orbitals of Sn seems stronger than with the vacant valence *d* orbitals of Pt and Ru. Following Pt_6Sn-ethanol-b, the dehydrogenation takes place with the H atom migrating to the neighbor Pt atom via the transition state structure (Pt_6Sn-ethanol-TS-b, $i349$ cm^{-1}) with an energy barrier of 0.57 eV that is considerably lower than that of C–H_α cleavage by 0.48 eV. As for another dehydrogenation process deriving from Pt_6Sn-ethanol-c, its transition state (Pt_6Sn-ethanol-TS-c) and the product complex (Pt_6Sn-ethanol-P-c) are very similar to the case of Pt_6Sn-ethanol-b dehydrogenation, like an energy barrier of 0.59 eV and dissociation energy of 0.50 eV. Thus, as ethanol is adsorbed on Sn site of Pt_6Sn the favorable dehydrogenation path is through the O–H cleavage instead of C_α-H_α cleavage. This result qualitatively follows the trend reported by Alcala et al., [15] in which the transition state energy to form CH_3CH_2O is lower by 0.12 eV than that to form CH_3CHOH on Pt_3Sn.

As discussed above, over Pt_7 and Pt_6Ru the favorable path for the first dehydrogenation reaction of ethanol is the C–H_α cleavage to form adsorbed CH_3CHO or CH_3CHOH (the respective E_a = 0.39 and 0.15 eV; E_{diss} = −0.94 and −0.38 eV). However, over Pt_6Sn the cleavage of hydroxyl has much lower energy barrier and dissociation energy than the C–H_α breaking (E_a: 0.57 vs. 1.05 eV; E_{diss}: 0.55 vs. 0.88 eV). These results indicate that whether the ethanol decomposition can proceed initially via the O–H or C–H_α bond cleavage depends on the alloyed metal in Pt-based bimetallic catalyst. In addition, one can see that, for the favorable reaction path of $Pt_6Sn+CH_3CH_2OH$, the hydrogen atom is preferably in contact with platinum atom, while the oxygen atom would rather contact with tin atom. This agrees with the theoretical result of the water decomposition on Pt_6Sn [13]. According to our calculated result shown in Fig. 3.8, not only are the adsorption energies of Pt_7-

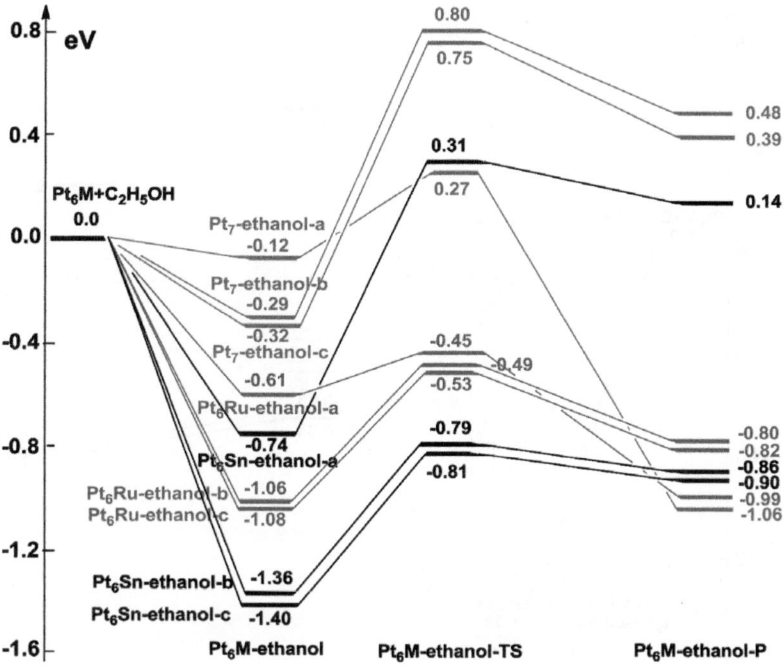

Figure 3.8 Reaction potential energy profiles of adsorption and dehydrogenation of CH_3CH_2OH on Pt_6M (M = Pt, Ru, Sn). Reprinted with permission from Xu and Wang [18]. Copyright © 2013 American Chemical Society.

ethanol-a, -b, and -c much smaller than those of Pt_6Sn-ethanol-a, -b and -c, but also Pt_7-ethanol-TS-a, the lowest energy transition state among the reaction of ethanol on Pt_7, is above reactants by 0.27 eV, whereas Pt_6Sn-ethanol-TS-b and -c are below the reactants by 0.79–0.81 eV. Thus, the adsorption and dehydrogenation reaction of ethanol on Sn site of Pt_6Sn is overwhelming. The electro-oxidation reaction of ethanol solution on the Pt-Sn catalyst would require certain atomic ratio of Sn to Pt in the metallic alloy so as to absorb more ethanol and water molecules. Recently, some experimental data show that the reactivity of ethanol electro-oxidation increases with the Sn atomic concentration of the Pt-Sn alloy increasing and the best ratio (Pt:Sn) is in the range of 2:1-1:1 [34–36].

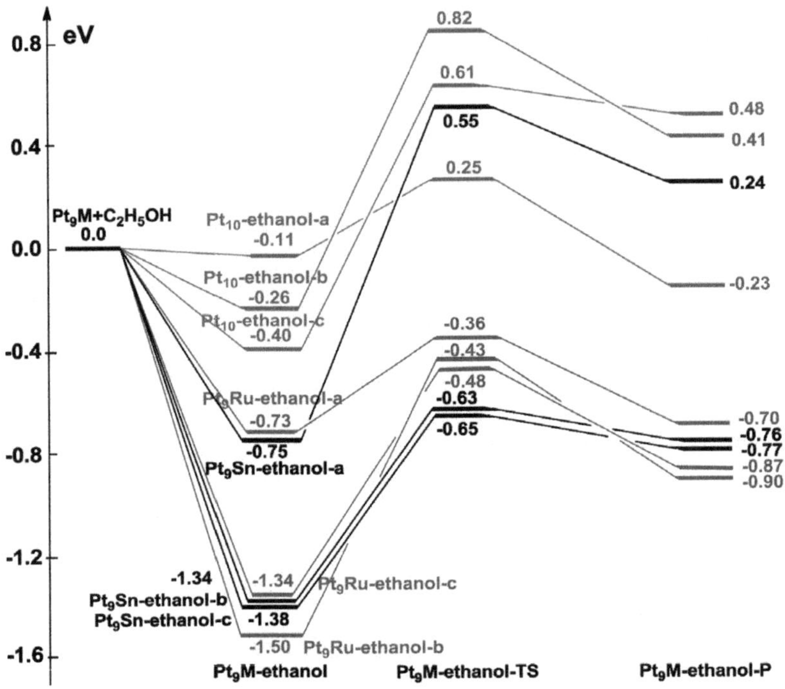

Figure 3.9 Reaction potential energy profiles of adsorption and dehydrogenation of CH_3CH_2OH on Pt_9M (M = Pt, Ru, Sn). Reprinted with permission from Xu and Wang [18]. Copyright © 2013 American Chemical Society.

3.4.2 CH_3CH_2OH Adsorption and Decomposition over Pt_9M Clusters (M = Pt, Sn, and Ru)

To discuss the layer effect, the second layer (Pt_3) was added to the relevant stationary points in Fig. 3.8, forming Pt_9M. The potential energy profiles in Fig. 3.9 were obtained with single-point energies for Pt_9M+ethanol. For the Pt_{10}+ CH_3CH_2OH reaction, the relative energies of the C–H_α adsorption complex (Pt_{10}-ethanol-a) and the transition state (Pt_{10}-ethanol-TS-a) are almost the same as those of Pt_7-ethanol-a and Pt_7-ethanol-TS-a, respectively. However, the dehydrogenation product (Pt_{10}-ethanol-P-a) is dramatically above Pt_7-ethanol-P-a by 0.83 eV. The relative energies of two hydroxyl adsorption reaction channels of this reaction system show difference from the one layer model, e.g., the adsorption through the hydroxyl of *cis*-ethanol is relatively more stable.

For the $Pt_9Ru+CH_3CH_2OH$ reaction, the adsorption energies of Pt_9Ru-ethanol-a, -b, and -c are more stable than those of Pt_6Ru-ethanol-a, -b, and -c by 0.12, 0.26, and 0.44 eV, respectively. The dehydrogenation products (Pt_9Ru-ethanol-P-a, -b, and -c) also have a small decrease in energy, and the energies of the dehydrogenation transition states have a small increase with a magnitude of less than 0.09 eV. Apparently, these changes are directly related to the fixed geometric parameters taken from the $Pt_6Ru + CH_3CH_2OH$ system.

For the energy prediction of the $Pt_9Sn + CH_3CH_2OH$ reaction, there is a good agreement with the $Pt_6Sn + CH_3CH_2OH$ reaction. The relative energies of the adsorption complexes (Pt_9Sn-ethanol-a, -b, and -c) are only 0.02 eV different from those in the $Pt_6Sn + CH_3CH_2OH$ reaction. The energies of the dehydrogenation transition states of the α-hydrogen and hydroxyl adsorption types are 0.24 and 0.16 eV, respectively, higher than those of the corresponding transition states in the $Pt_6Sn + CH_3CH_2OH$ reaction. The energy increase of the dehydrogenation products is also less than 0.13 eV compared to the products of $Pt_6Sn + CH_3CH_2OH$.

It is apparent that the two layer Pt-based cluster model for the ethanol decomposition may not be accurate yet, but the results have the same trend as those of the one layer Pt-based cluster model, especially for the dominant reaction channels. Thereforc, this consistency can make us believe that thc calculaled result, based on the one-layer model, may be reliable to clarify the electro oxidation mechanism of ethanol.

3.4.3 Kinetics of Adsorption and Decomposition of CH_3CH_2OH

The above adsorption and dehydrogenation mechanisms of ethanol on the surface of Pt-based alloy can be represented by the reaction scheme:

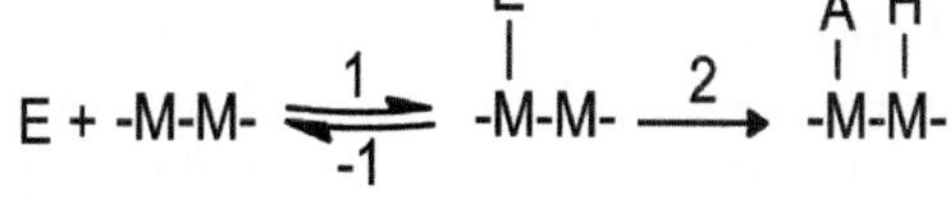

E is ethanol and –M– is the surface of the alloy, while A– and H– are the decomposition products of ethanol. On the basis of the steady state approximation, the overall rate constant can be expressed as

$$k = \frac{k_2 k_1}{k_{-1} + k_2} = \frac{K k_2}{1 + k_2/k_{-1}} \tag{3.1}$$

where k_1 and k_{-1} are the adsorption and desorption reaction rate constants, and k_2 is the reaction rate constant for the decomposition; K $(= k_1/k_{-1})$ is the adsorption equilibrium constant. According to the transition state rate theory, the decomposition reaction rate constant is formulated as

$$k_2 = L\frac{\kappa T}{h}\frac{f^{\neq}}{f_{\mathrm{M}}}\exp\left(-\frac{\Delta E}{RT}\right) \tag{3.2}$$

and the adsorption equilibrium constant is

$$K = \frac{f_{\mathrm{ME}}}{f_{\mathrm{E}} f_{\mathrm{M}}}\exp\left(-\frac{\Delta H}{RT}\right) \tag{3.3}$$

In equations (3.2) and (3.3), κ is Boltzman constant and h is Plank constant; ΔE is the potential barrier of dehydrogenation reaction; ΔH is the adsorption energy of adsorption process; $f^{\neq}$ is the partition function of transition state; f_{E}, f_{M}, and f_{ME} are the partition functions of ethanol, metal, and metal-ethanol adsorption complex, respectively; and L is the number of adsorption sites per area unit. For the pure Pt metallic surface, L is evaluated to be 5.79×10^{14} cm^{-2}, while the adsorption sites at the atomic ratio of 2:1 for Pt:Ru and Pt:Sn are evaluated to be 1.71×10^{14} and 1.60×10^{14} cm^{-2}, respectively. The partition function of ethanol is approximated as the sum of the translation, rotation, vibration, and electron contributions, while the partition functions of the Pt-based cluster and their adsorption complexes as well as transition states only involve both electronic and vibrational contributions because of their solid property.

In the light of the above discussion on the adsorption and dehydrogenation reaction mechanisms, the rate constants are calculated only for the most favorable reaction channels, such as α-hydrogen adsorption for Pt_7+CH_3CH_2OH, hydroxyl adsorption for Pt_6Sn+CH_3CH_2OH, and both α-hydrogen adsorption and hydroxyl adsorption for Pt_6Ru+CH_3CH_2OH. Figure 3.10 shows the predicted rate constants of three reaction systems in the temperature range

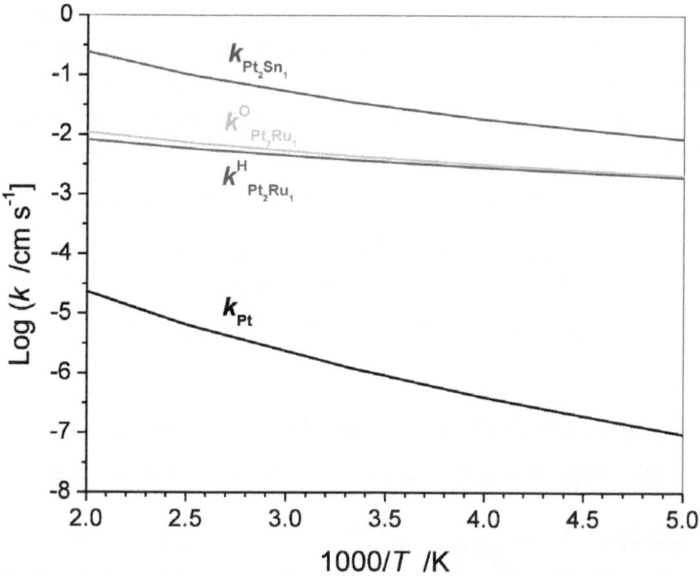

Figure 3.10 Predicted rate constants of adsorption and dehydrogenation of CH_3CH_2OH on Pt-M catalysts (M = Pt, Ru, Sn) versus reciprocal temperature. Reprinted with permission from Xu and Wang [18]. Copyright © 2013 American Chemical Society.

from 200 K to 500 K. The rate constants of the Pt_7+CH_3CH_2OH, Pt_6Ru+CH_3CH_2OH, and Pt_6Sn+CH_3CH_2OH systems are denoted by k_{Pt}, k_{Pt2Ru1}, and k_{Pt2Sn1}, respectively. It can be seen that all of the reaction rate constants of ethanol with the metallic catalysts appear positive temperature dependent, which implies that their reactivity increases with the temperature increasing in spite of the weak temperature dependence of k_{Pt2Ru1}. Apparently, k_{Pt2Sn1} is the highest while k_{Pt} is the lowest; however, k_{Pt2Ru1} is less than the former but much higher than the latter. For example, at 300 K k_{Pt2Sn1} is 4.3 and 2.9×10^4 times greater than k_{Pt2Ru1} and k_{Pt}, respectively. Moreover, at 400 K k_{Pt2Sn1} is 7.7 and 1.7×10^4 times higher than k_{Pt2Ru1} and k_{Pt}, respectively. In addition, one can see that for k_{Pt2Ru1} the rate constant line of the hydroxyl adsorption type reaction is slightly greater than that of the α-hydrogen one. For example, k^{O}_{Pt2Ru1} is over k^{H}_{Pt2Ru1} by 1.3 and 1.4 times at 300 and 400 K, respectively. This shows in kinetics that the adsorption and dehydrogenation reaction of ethanol on

Ru site of Pt-Ru and Sn site of Pt-Sn occurs favorably through the hydroxyl adsorption. For practical application, these predicted rate constants in units of cm·s^{-1} are fitted by three parameters Arrhenius expressions as follows in the temperature range from 200 K to 500 K.

$$
\begin{aligned}
k_{\mathrm{Pt}} &= 1.38 \times 10^{-18} T^{4.99} \exp(-299/\mathrm{T}) \\
k^{\mathrm{H}}_{\mathrm{Pt2Ru1}} &= 1.61 \times 10^{-6} \mathrm{T}^{1.39} \exp(-53/\mathrm{T}) \\
k^{\mathrm{O}}_{\mathrm{Pt2Ru1}} &= 1.95 \times 10^{-7} \mathrm{T}^{1.76} \exp(-15/\mathrm{T}) \\
k_{\mathrm{Pt2Sn1}} &= 2.22 \times 10^{-13} \mathrm{T}^{4.39} \exp(224/\mathrm{T})
\end{aligned}
$$

3.5 Alloying Effect on the Adsorption Energy, Energy Barrier, and Dissociation Energy

To discuss the ligand effect, adsorption and the consequent dehydration for ethanol op the M site as well as Pt site of Pt_3M (M = Pt, Ru, Sn, Re, Rh, and Pd) were extensively investigated [17], as shown in Fig. 3.11 for Pt_3Re. As discussed above, the alloyed atoms M of PtM may affect properties of the adsorption and dehydrogenation of α-hydrogen and hydroxyl on Pt, which are summarized in Fig. 3.12. It can be seen that from Fig. 3.12A, compared with Pt_4, the alloying of Sn enhances the adsorption on Pt site of Pt_3Sn for both the α-H and hydroxyl of ethanol, which favors dehydrogenation rate. The adsorption of the α-H on the Sn site of Pt_3Sn is unstable, and the adsorption of the hydroxyl group on the Sn site of Pt_3Sn is similar to that on the Pt site. Re alloying slightly enhances the adsorption of the hydroxyl on the Pt site, *but it does not favor the adsorption of the* α-H. Ru and Rh decrease both adsorptions on the Pt site of PtRu/Rh; however, alloying of Pd does not affect the adsorptions on the Pt site of PtPd.

According to Fig. 3.12B, for all of the investigated metals the dehydrogenation barriers for the α-H adsorption complex (bottom) are much lower than those for the hydroxyl adsorption complexes (top). For the kinetically favorable α-H path, as compared with Pt_4 only Pd has a similar barrier, and all of others have higher barrier than Pt_4. *Indeed, neither Pt nor Sn site of PtSn can be more active to*

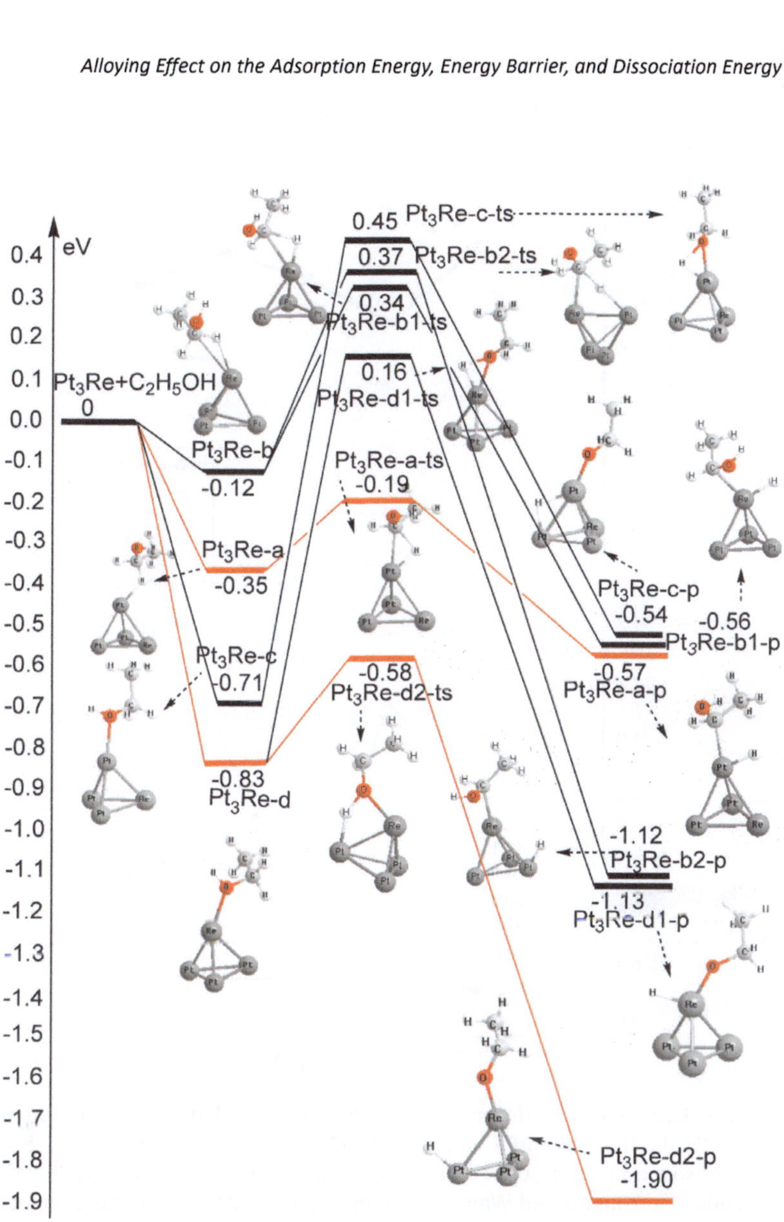

Figure 3.11 Potential energy profile for the adsorption and dehydrogenation of CH_3CH_2OH on Pt_3Re, with B3PW91 including ZPE. The top two lines refer to the α-H pathway on the Pt site (Pt_3Re-a, red) and the Re site (Pt_3Re-b, black), while the bottom ones to the hydroxyl pathway on the Re site (Pt_3Re-d, red) and the Pt site (Pt_3Re-c, black). Reprinted with permission from Xu and Wang [17]. Copyright © 2011 American Chemical Society.

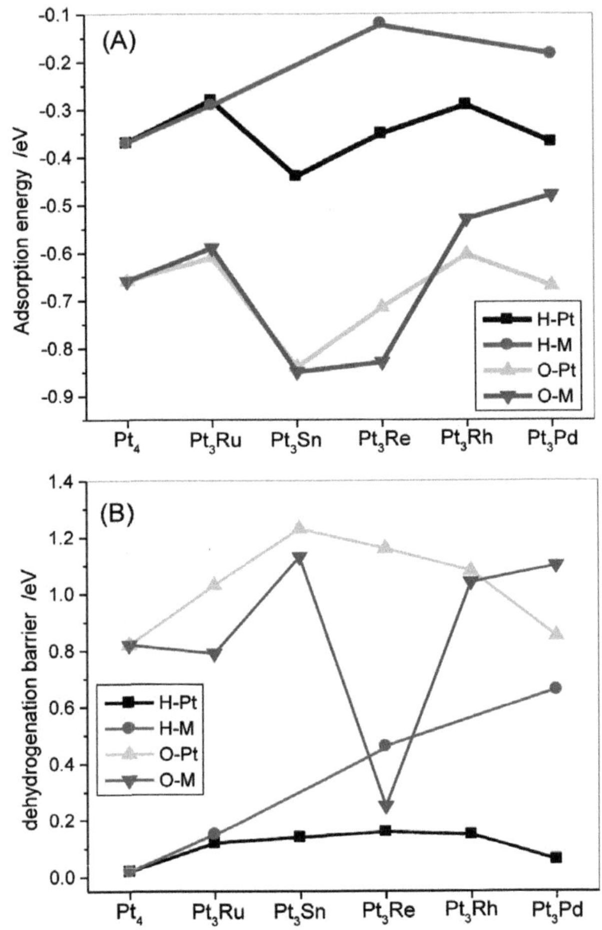

Figure 3.12 Ligand effects on the adsorption and dehydrogenation of C_2H_5OH on the bimetallic clusters, Pt_3M (M = Pt, Ru, Sn, Re, Rh, Pd). (A) Adsorption energies. (B) Dehydrogenation barriers. Reprinted with permission from Xu and Wang [17]. Copyright © 2011 American Chemical Society.

the dehydration of ethanol than pure Pt. However, Re site of PtRe can only better catalyze the hydroxyl path. The effect of metal alloying in Pt will be further thoroughly explored with a larger cluster model.

In order to analyze the interaction of ethanol molecule with the bimetallic catalysts, the highest occupied molecular orbital (HOMO)

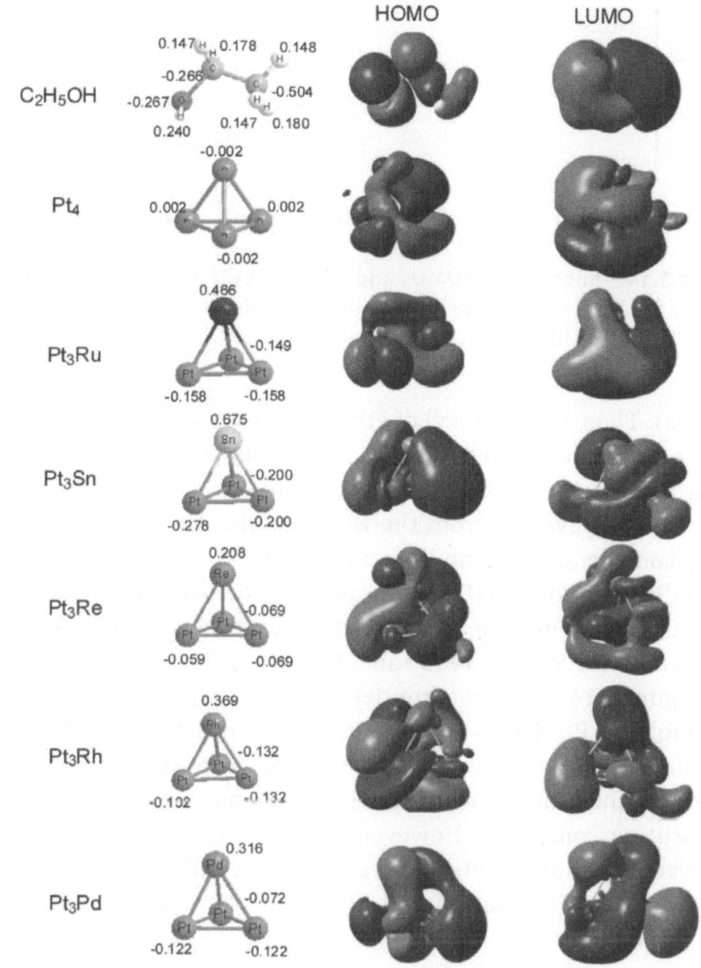

Figure 3.13 Frontier molecular orbitals and Mulliken atomic charges of CH_3CH_2OH and Pt_3M (M = Pt, Ru, Sn, Re, Rh, Pd) Reprinted with permission from Xu and Wang [17]. Copyright © 2011 American Chemical Society.

and the lowest unoccupied molecular orbital (LUMO) are drawn in Fig. 3.13 with Mulliken atomic charges. It is known that the hydroxyl of ethanol binds to Pt through the lone electron pair of the oxygen. Figure 3.7 reveals that HOMO's phase of ethanol matches well with the LUMO of all of investigated Pt_3M (M = Pt, Ru, Sn, Re,

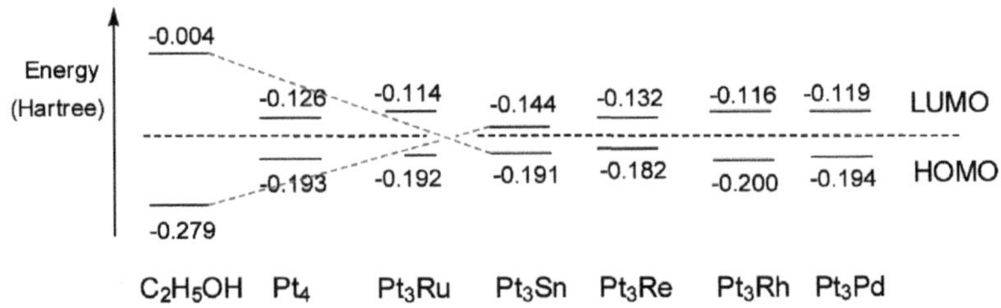

Figure 3.14 Energies of HOMOs and LUMOs of CH_3CH_2OH and Pt_3M (M = Pt, Ru, Sn, Re, Rh, Pd). Reprinted with permission from Xu and Wang [17] Copyright © 2011 American Chemical Society.

Rh, Pd). Thus, the hydroxyl of ethanol has stable adsorptions on all of the investigated metallic sites. The unstable adsorptions for the α-hydrogen of ethanol on the sites of Sn of Pt_3Sn and Rh site of Pt_3Rh can also be revealed from the HOMO plots. Although the HOMOs have considerable distribution adjacent to Pt sites, Ru of Pt_3Ru, Re of Pt_3Re, and Pd of Pt_3Pd, minor electron density on the Sn and Rh sites is found. Thus, the α-hydrogen adsorption complexes were not located on the Sn site of PtSn and Rh site of PtRh.

Figure 3.14 shows the frontier molecular orbital energy levels of ethanol and Pt_3M (M = Pt, Ru, Sn, Re, Rh, Pd). It can be seen that, because the bimetallic models have very close HOMOs and LUMOs energies, they should have similar interactions with ethanol to form adsorption complexes. However, the best interaction may happen between ethanol and Pt_3Sn since the energy differences of E(LUMO, Pt_3Sn)-E(HOMO, ethanol) and E(LUMO, ethanol)-E(HOMO, Pt_3Sn) are the smallest.

3.6 Conclusion

Although the cluster models are unable to simulate bulk surface properties, they are suitable for simulating the properties of nanoscale particle catalysts. Moreover, since the major objective of the present investigation is to compare the catalytic activity of various PtM bimetallic nanoparticles, the trial error due to the cluster model and omission of solvent environment may be limited to some degree. Water adsorption on the Sn site of Pt_nSn ($n = 2$ and 3) is weaker than those on the Pt site of Pt_n ($n = 3$ and 4) and the Ru site of Pt_nRu ($n = 2$ and 3), while water adsorptions on the central Sn atom of Pt_6Sn and Pt_9Sn are enhanced so significantly that they are stronger than those on the central Pt and Ru atoms of Pt_nM ($n = 6$ and 9; M = Pt and Ru). For all of the three cluster models, energy barrier (E_a) for the dissociation of adsorbed water over Sn is lower than over Ru and Pt atoms (e.g., E_a: 0.78 vs. 0.96 and 1.07 eV for Pt_9M). It is interesting to note that the dissociation energy on Sn site is also the lowest (E_{diss}: 0.44 vs. 0.61 and 0.67eV). The results show that from both kinetic and thermodynamic viewpoints Sn is more active to water decomposition than pure Pt and the other investigated PtM alloys except for PtRe, and the kinetic activity for water dissociation increases in the sequence of Cu $<$Pd $<$Rh $<$Pt $<$Ru $<$Sn $<$Re, which well supports the assumption of the bifunctional mechanism that Sn site accelerates the dissociation of H_2O. However, *although the energy barrier for water decomposition on the Re site of the PtRe is only slightly lower than that on the Sn site, the adsorption energy and dissociation energy on the Re site are considerably more favorable than those on the Sn site of PtSn.* The adsorptions of ethanol on Pt and M sites of PtM via the α-hydrogen were well established. Although the α-hydrogen adsorption on Pt site is weaker than the hydroxyl of ethanol, the potential energy profiles show that the dehydrogenation via the α-H path has much lower energy barrier than that via the hydroxyl path. Generally for the α-hydrogen path the adsorption is a rate-determining step because of rather low dehydrogenation barrier for the α-hydrogen adsorption complex (thermodynamic control), while the hydroxyl path is determined by its dehydrogenation step (kinetic control). The effects of alloyed metal on the catalysis activity of Pt for ethanol partial oxidation, including adsorption energy, energy barrier, and electronic structure were discussed. Alloying of Sn enhances the adsorption of ethanol on Pt site through the α-H, as well as those on both Pt and Sn sites through the hydroxyl. However, the Re alloying does not enhance the former adsorption although it enhances the latter one on both Pt and Re sites.

References

1. Song, S., and Tsiakaras, P. (2006). Recent progress in direct ethanol proton exchangemembrane fuel cells (DE-PEMFCs), *Appl. Catal. B: Environ.*, **63**, 187–193.
2. Jiang, L., Sun, G., Sun, S., Liu, J., Tang, S., Li, H., Zhou, B., and Xin, Q. (2005). Structure and chemical composition of supported Pt-Sn electrocatalysts for ethanol oxidation, *Electrchim. Acta*, **50**, 5384–5389.
3. Vigier, F., Coutanceau, C., Perrard, A., Belgsir, E. M., and Lamy, C. (2004). Development of anode catalysts for a direct ethanol fuel cell, *J. Appl. Electrochem.*, **34**, 439–446.
4. Vigier, F., Coutanceau, C., Hahn, F., Belgsir, E. M., and Lamy, C. (2004). On the mechanism of ethanol electro-oxidation on Pt and PtSn catalysts: electrochemical and in situ IR reflectance spectroscopy studies, *J. Electroanal. Chem.*, **563**, 81–89.
5. Mann, J., Yao, N., and Bocarsly, A. B. (2006). Characterization and analysis of new catalysts for a direct ethanol fuel cell, *Langmuir*, **22**, 10432–10436.
6. Rousseau, S., Coutanceau, C., Lamy, C., and Leger, J.-M. (2006). Direct ethanol fuel cell (DEFC): electric performances and reaction products distribution under operating conditions with different platinum-based anodes, *J. Power Sources*, **158**, 18–24.
7. Lamy, C., Rousseau, S., Belgsir, E. M., Coutanceau, C., and Leger, J.-M. (2004). Recent progress in the direct ethanol fuel cell: development of new platinum-tin electrocatalysts, *Electrchim. Acta*, **49**, 3901–3908.
8. Song, S. Q., Zhou, W. J., Zhou, Z. H., Jiang, L. H., Sun, G. Q., Xin, Q., Leontidis, V., Kontou, S., and Tsiakaras, P. (2005). Direct etahnol PEM fuel cells: the case of platinum based anods, *Int. J. Hydrogen Engergy*, **30**, 995–1001.
9. Zhu, M., Sun, G., and Xin, Q. (2009). Effect of alloying degree in PtSn catalyst on the catalytic behavior for ethanol electro-oxidation, *Electrochim. Acta*, **54**, 1511–1518.
10. Watanabe, M., and Motoo, S. (1975). Electrocatalysis by ad-atoms, *J. Electroanal. Chem.*, **60**, 275–283.
11. Iwasita, T., and Pastor, E. (1994). A DEMS and FTIR spectroscopic investigation of adsorbed ethanol on polycrystalline platinum, *Electrochim. Acta*, **39**, 531.
12. Lux, K. W. C., and Elton J. (2006). Lanthanide-platinum intermetallic compounds as anode electrocatalysts for direct ethanol PEM fuel cells.

II. Performance of LnPt2 (Ln = Ce, Pr) nanopowders in an operating PEM fuel cell, *J. Electrochem. Soc.*, **153**, A1139–A1147.

13. Wang, Y., Mi, Y., Redmon, N., and Holiday, J. (2010). Understanding electrocatalytic activity enhancement of bimetallic particles to ethanol electro-oxidation. 1: Water Adsorption and Decomposition on Pt_nM (n = 2, 3, and 9; M = Pt, Ru, and Sn), *J. Phys. Chem. C*, **114**, 317–326.
14. Alcala, R., Mavrikakis, M., and Dumesic, J. A. (2003). DFT studies for cleavage of C–C and C–O bonds in surface species from ethanol on Pt(111), *J. Catal.*, **218**, 178–190.
15. Alcala, R., Shabaker, J. W., Huber, G. W., Sanchez-Castillo, M. A., and Dumesic, J. A. (2005). Experimental and DFT Studies of the conversion of ethanol and acetic acid on PtSn-based catalysts, *J. Phys. Chem. B*, **109**, 2074–2085.
16. Wang, H.-F., and Liu, Z.-P. (2007). Selectivity of direct ethanol fuel cell dictated by a unique partial oxidation channel, *J. Phys. Chem. C*, **111**, 12157–12160.
17. Xu, Z.-F., and Wang, Y. (2011). Effects of alloyed metal on the catalysis activity of Pt for ethanol partial oxidation: adsorption and dehydrogenation on Pt_3M (M = Pt, Ru, Sn, Re, Rh, and Pd), *J. Phys. Chem. C*, **115**, 20565–20571.
18. Xu, Z., and Wang, Y. W. (2013). Understanding electrocatalytic activity enhancement of bimetallic particles to ethanol electro-oxidation: ethanol adsorption and decomposition on Pt_nM (n = 6 and 9; M = Pt, Ru, and Sn), chap. 8 in *Applications of Molecular Modeling to Challenges in Clean Energy*, ed. Fitzgerald, G., Govind, N. ACS Symposium Series. New York: Americal Chemical Society, **1113**, pp. 135–151.
19. Ishikawa, Y., Liao, M.-S., and Cabrera, C. R. (2002). Energetics of H_2O dissociation and COads+OHads reaction on a series of Pt-M mixed metal clusters. A relativistic density-functional study, *Surf. Sci.*, **513**, 98–110.
20. Kandoi, S. G., A. A., Grabow, L. C., Dumesic, J. A., and Mavrikakis, M. (2004). Why Au and Cu are more selective than Pt for preferential oxidation of CO at low temperature, *Catal. Lett.*, **93**, 93–100.
21. Meng, S., Wang, E. G., and Gao, S. (2004). Water adsorption on metal surfaces. A general picture from density functional theory studies, *Phys. Rev. B*, **69**, 195404/1-195404/13.
22. Phatak, A. A., Delgass, W. N., Ribeiro, F. H., and Schneider, W. F. (2009). Density Functional Theory Comparison of Water Dissociation Steps on Cu, Au, Ni, Pd, and Pt, *J. Phys. Chem. C*, **113**, 7269–7276.

23. Parreira, R. L. T., Caramori, G. F., Galembeck, S. E., and Huguenin, F. (2008). The nature of the interactions between Pt_4 cluster and the adsorbates •H, •OH, and H_2O, *J. Phys. Chem. A*, **112**, 11731–11743.
24. Michaelides, A., and Hu, P. (2001). Catalytic water formation on platinum: a first-principles study, *J. Am. Chem. Soc.*, **123**, 4235–4242.
25. Grabo, L. C., Gokhale, A. A., Evans, S. T., Dumesic, J. A., and Mavrikakis, M. (2008). Mechanism of the water gas shift reaction on Pt: first principles, experiments, and microkinetic modeling, *J. Phys. Chem. C*, **112**, 4608–4617.
26. Wang, G.-C., Tao, S.-X., and Bu, X.-H. (2006). A systematic theoretical study of water dissociation on clean and oxygen-preadsorbed transition metals, *J. Catal.*, **244**, 10–16.
27. Cao, Y., and Chen, Zhao-Xu (2006). Theoretical studies on the adsorption and decomposition of H_2O on Pd(111) surface, *Surf. Sci.*, 4572–4583.
28. Greeley, J., and Mavrikakis, M. (2004). Competitive paths for methaol decomposition on Pt(111), *J. Am. Chem. Soc.*, **126**, 3910–3919.
29. Desai, S. K., Neurock, M., and Kourtakis, K. (2002). A periodic density functional theory study of the dehydrogenation of methanol over Pt(111), *J. Phys. Chem. B*, **106**, 2559.
30. Yang, M.-M., Bao, X.-H., and Li, W.-X. (2007). First principle study of ethanol adsorption and formation of hydrogen bond on Rh(111) surface, *J. Phys. Chem. C*, **111**, 7403–7410.
31. Alcala, R., Shabaker, J. W., Huber, G. W., Sanchez-Castillo, M. A., and Dumesic, J. A. (2005). Experimental and DFT studies of the conversion of ethanol and acetic acid on PtSn-based catalysts, *J. Phys. Chem. B*, **109**, 2074–2085.
32. Vigier, F., Rousseau, S., Coutanceau, C., Leger, J.-M., and Lamy, C. (2006). Electrocatalysis for the direct alcohol fuel cell, *Top. Catal.*, **40**, 111–121.
33. Lee, A. F., Gawthrope, D. E., Hart, N. J., and Wilson, K. (2004). A fast XPS study of the surface chemistry of ethanol over Pt(111), *Surf. Sci.*, **548**, 200–208.
34. Zhou, W. J., Song, S. Q., Li, W. Z., Suna, G. Q., Xin, Q., Kontou, S., Poulianitis, K., and Tsiakaras, P. (2004). Pt-based anode catalysts for direct ethanol fuel cells, *Solid State Ionics*, **175**, 797–803.
35. Zhou, W. J., Song, S. Q., Li, W. Z., Zhou, Z. H., Suna, G. Q., Xin, Q., Douvartzides, S., and Tsiakaras, P. (2005). Direct ethanol fuel cells based

on PtSn anodes: the effect of Sn content on the fuel cell performance, *J. Power Sources*, **140**, 50–58.

36. Tsiakaras, P. E. (2007). PtM/C (M = Sn, Ru, Pd, W) based anode direct ethanol-PEMFCs: structural characteristics and cell performance, *J. Power Sources*, **171**, 107–112.

Chapter 4

Theoretical Aspects of Gold Nanocatalyst for Ethanol and Glucose Oxidation

Takayoshi Ishimoto and Michihisa Koyama

INAMORI Frontier Research Center and International Institute for Carbon-Neutral Energy Research, Kyushu University, Motooka 744, Nishi-ku, Fukuoka 819-0395, Japan

koyama@ifrc.kyushu-u.ac.jp

Ethanol and other biofuels will play an important role in future low-carbon society. The direct use of such compounds in fuel cell is one of the promising options to realize high-efficiency utilization of biofuels. In this chapter, we focus on the mechanistic aspects of ethanol and glucose oxidation in alkaline environment to provide insights into the direct use of biofuels in fuel cells. Au is selected as a promising catalyst alternative to Pt. To understand the catalytic activity of Au in alkaline environment, we first analyzed by using density functional theory the surface state of Au in alkaline solution and the oxidation reaction of ethanol and glucose to understand key factors in their electro-oxidation process in direct alkaline fuel cell. From the calculated adsorption energy of OH^- in alkaline solution, high OH^- coverage of the Au surface in alkaline solution is indicated. As a first step to analyze the oxidation reaction mechanism of ethanol, its state in alkaline solution is investigated. It

Nanomaterials for Direct Alcohol Fuel Cell
Edited by Yixuan Wang

ISBN 978-981-4669-00-9 (Hardcover), 978-981-4669-01-6 (eBook)
www.panstanford.com

was confirmed that the ethanol dissociates to be deprotonated form under high pH condition. The oxidation process is studied assuming the deprotonated form of ethanol on OH-covered Au surface. We followed the same way to tackle the mechanism of glucose oxidation on Au catalyst in alkaline environment. We finally analyzed the effect of support materials on glucose oxidation as a measure for further improvement of activity of gold catalyst. The factors influencing the high catalytic activity of gold are discussed.

4.1 General Introduction

The selective oxidation of biofuels, such as ethanol and glucose, will be a sustainable, environmentally important process in future low-carbon society [1–5]. To achieve the high efficiency utilization of biofuels, alkaline fuel cells (AFCs) is one of the promising options [6–8]. Although the platinum-based catalysts are mainly used for ethanol oxidation in low-temperature fuel cells [9–18], one of the serious problems of platinum-based catalysts for biofuel is the lack catalytic activity due to CO poisoning [19–21]. It is thus necessary to develop high efficient catalysts for biofuels. Recently, Au has attracted interests as electrocatalyst for biofuels in alkaline environment [22–41]. For example, Koper et al. found the promoting effect of adsorbed CO on the oxidation of alcohols on a Au catalyst [26]. This result suggests that it is possible to avoid CO poisoning by Au catalyst. They also found that the current density for ethanol oxidation drastically increases at high pH region [27].

In this sense, the Au catalyst in alkaline solution plays an important role for highly efficient oxidation of biofuels. To understand the catalytic activity in alkaline solution, the adsorption properties of hydroxyl on the Au surface should be clarified as a first step. The oxidation reaction of ethanol and glucose proceeds on the intrinsic Au surface in alkaline solution. In this chapter, we summarize our recent activities on the oxidation reaction of ethanol and glucose on Au catalyst in direct alkaline fuel cell by using density functional theory (DFT) calculation. Computational details based on DFT are shown in Section 4.2. First, we show Au surface structure in alkaline solution in Section 4.3. Oxidation reaction of ethanol and glucose on Au surface are summarized in Sections 4.4 and 4.5, respectively. Finally, we introduce in Section 4.6 the effect of support materials. More detailed backgrounds of each topic are shown in each section. We describe summary and perspectives in Section 4.7.

4.2 Computational Details

All calculations were performed under the generalized gradient approximation (GGA) with Perdew-Wang 1991 exchange and correlation functional (PW91) [42, 43] as implemented in the $DMol^3$ package [44, 45]. Double numerical atomic basis sets augmented with polarization function (DNP) were used to describe the valence electrons, and the core electrons were represented by effective core potentials (ECP). Solvent effects in water were estimated using the conductor-like screening model (COSMO) [46]. The transition state calculations were performed using synchronous transit methods, the linear synchronous transit (LST), and quadratic synchronous transit (QST) [47, 48]. To obtain atomic charges, Hirshfeld approach [49] was used.

4.3 Au Surface Structure in Alkaline Solution

4.3.1 Introduction

Several studies have been reported to theoretically understand the stability and reactivity of Au surface [50–53]. While the majority is focusing on the acidic environment, Pessoa et al. pointed out that in alkaline solution the OH adsorption is the first step of electro-oxidation reactions and may act as catalytic promoter or adsorption preference [50]. Therefore, we have investigated the three aspects of the effects of alkaline environment on the catalytic process stepwise: (i) effect on the states of gold surface, (ii) effect on the states of reactants, and (iii) effect of OH^- species as oxidant. In this section, we describe the analyses on the states of gold surface and reactants in alkaline environment [54, 55].

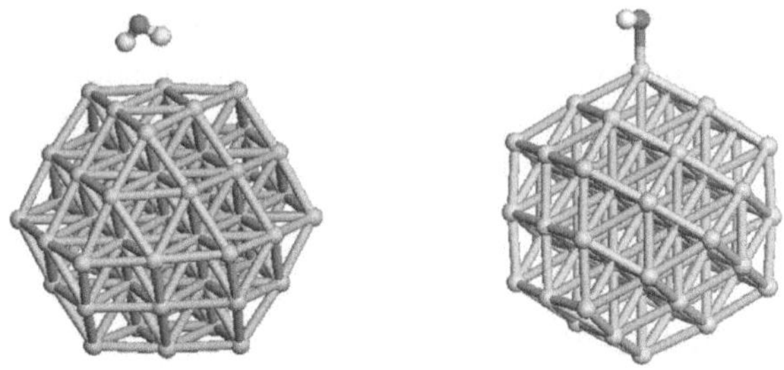

Figure 4.1 Adsorption energy and structure of H_2O and OH^- on Au_{55} cluster.

4.3.2 H_2O and OH^- Adsorption Energy on Au

To understand the surface condition of Au in alkaline solution, we calculated the adsorption energy of H_2O and OH^- molecules on Au_{55} cluster. The optimized structures of H_2O and OH^- adsorbed Au_{55} cluster are shown in Fig. 4.1.

Because the adsorption energies of H_2O and OH^- were −84.5 and −211.5 kJ/mol, respectively, the OH^- adsorption on Au surface is dominant. The adsorption energy of multiple OH^- species on Au_{55} cluster is shown in Fig. 4.2. Adsorption energy in Fig. 4.2 means average of adsorption energy for one OH^-. When the number of surface OH^- increases, the adsorption energy of OH^- decreases. However, these adsorption energies of OH^- molecules are still smaller than that of H_2O molecule. This result indicates that the Au surface in alkaline solution will be covered by OH^- molecules.

We next calculated atomic charge population of OH^- and Au_{55} cluster to understand the electronic structure of adsorbed OH^- on Au surface by using the Au_{55} cluster with adsorbed OH^- or OH species (model in Fig. 4.1). When the OH^- adsorbed on Au_{55} cluster, the electron in OH^- transferred to Au_{55} cluster. As a result, the atomic charge population of adsorbed OH^- is close to OH. In addition, the transferred electron from OH^- to Au catalyst is used as an electric power during the AFC operation condition. In this sense, we assumed the neutral cluster model for the OH^- adsorbed Au cluster.

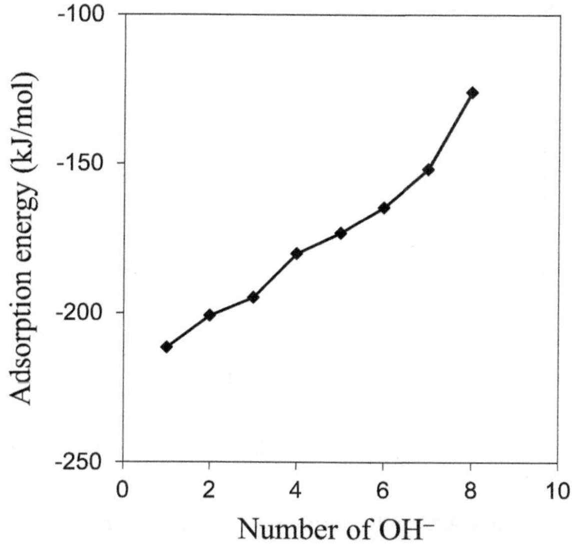

Figure 4.2 OH^- adsorption energy on Au_{55} cluster.

4.3.3 Adsorption Species on Au Surface in Alkaline Solution

We discuss about the stable structure of ethanol, which is an example of biofuels, in alkaline solution. Equilibrium reaction of ethanol in alkaline environment is the following.

$$CH_3CH_2OH + OH^- \rightleftarrows CH_3CH_2O^- + H_2O \qquad (4.1)$$

The free energy of Eq. 4.1 at 298 K was calculated to be −36.2 kJ/mol. From the obtained value, we then calculated the dissociation ratio of CH_3CH_2OH at different pH as shown in Fig. 4.3. When pH is above 11, ethanol is estimated to dissociate into $CH_3CH_2O^-$. This would partly explain the high reactivity of alcohol in alkaline environment because the initial hydrogen abstraction from ethanol is a key step when the platinum catalyst is used in acidic environment.

We then analyzed the adsorption properties of major chemical species in reaction environment. We assumed H_2O, OH^-, CH_3CH_2OH, $CH_3CH_2O^-$, CH_3CHO, and CH_3COO^- as chemical species. Table 4.1 shows the adsorption energy of species on Au_{13} cluster model.

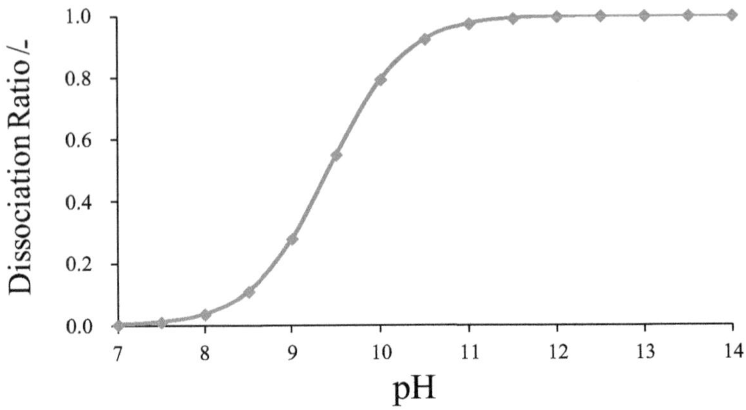

Figure 4.3 Dissociation ratio of ethanol in alkaline environment.

Table 4.1 Adsorption energy on Au_{13} cluster model

Adsobates	Adsorption energy (kJ/mol)
H_2O	14.4
OH^-	−173.8
CH_3CH_2OH	14.5
$CH_3CH_2O^-$	−143.6
CH_3CHO	0.9
CH_3COO^-	−74.8

Among the studied species, it was found that OH^-, $CH_3CH_2O^-$, and CH_3COO^- are stabilized on the Au surface.

We then estimated the surface coverage of each species on the Au surface assuming the Langmuir's adsorption model. In the estimation, we assumed that CH_3CH_2OH and $CH_3CH_2O^-$ are in equilibrium of Eq. 4.1 and summed concentrations of those two species are assumed to be 0.5 mol/L. We also assumed the concentrations of CH_3CHO and CH_3COO^- as 0.5 mol/L for simplicity. Figure 4.4 shows the estimated surface coverage. We can see that OH^- dominantly occupies the surface adsorption site at pH range above 11. $CH_3CH_2O^-$, showing second strong adsorption energy, occupies a certain portion of adsorption sites at lower pH.

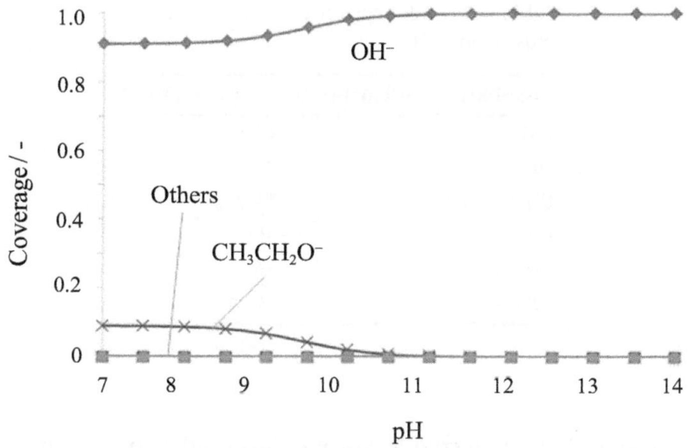

Figure 4.4 Surface coverage species on Au in alkaline environment.

Table 4.2 Adsorption energy on Au_{13}–OH cluster model

Adsobates	Adsorption energy (kJ/mol)
H_2O	−5.0
OH^-	−92.5
CH_3CH_2OH	−3.1
$CH_3CH_2O^-$	−48.3
CH_3CHO	17.8
CH_3COO^-	−22.0

We next analyzed a second layer adsorption on the Au surface covered by OH^-. We used a simplified model having single OH^- on Au_{13} surface to investigate the adsorption of other chemical species. Table 4.2 shows the adsorption energies onto Au_{13}–OH cluster model. In the case of second layered OH^- adsorption, we observed the following reaction.

$$CH_3CH_2OH + OH^- \rightleftarrows CH_3CH_2O^- + H_2O \tag{4.2}$$

This indicates that the Au surface is covered by OH^-, however surface OH^- reacts with OH^- in solution to form surface O as reported in previous study [56].

Table 4.3 Adsorption energy on Au_{13}–O cluster model

Adsobates	Adsorption energy (kJ/mol)
H_2O	−11.1
OH^-	−36.2
CH_3CH_2OH	−14.1
$CH_3CH_2O^-$	−8.0
CH_3CHO	36.0
CH_3COO^-	−3.0

We therefore analyzed a second layer adsorption on the Au surface covered by O. We used a simplified model having single O on Au_{13} surface to investigate the adsorption of other chemical species. Table 4.3 shows the adsorption energies onto Au_{13}–O cluster model. Compared with adsorption onto bare Au_{13} and OH-covered Au_{13} cluster models, adsorption of $CH_3CH_2O^-$, potential initial reactant, was found to be weak, indicating a relatively low activity against the oxidation of ethanol.

From those analyses, it was found that (i) OH^- strongly adsorbs on Au surface to have high coverage and (ii) adsorbed OH^- reacts with liquid phase OH^- to form adsorbed O and H_2O. However, we may need further analyses to conclusively discuss the surface states of Au because the present calculations do not consider the electric potential, interadsorbate interactions, and the influence of particle size.

4.3.4 Summary

We estimated the Au surface state in alkaline solution. The adsorption energy of OH^- was larger than that of H_2O, indicating that the Au surface in alkaline solution is covered by OH^-. It was confirmed that the ethanol dissociates to be $CH_3CH_2O^-$ under high pH condition. From the analysis of adsorption on Au surface, it was found that the Au surface under practical operation condition would be covered by OH^- or O while we still need further analyses.

4.4 Oxidation Reaction of Ethanol on Au Catalyst

4.4.1 Introduction

Concerning the oxidation mechanism of ethanol, which is one of the biofuels, there are many experimental and theoretical studies [14–18, 26–28, 57]. Many of these studies are ethanol oxidation on platinum-based catalysts in acidic condition [14–18]. It is expected that the oxidation mechanism in alkaline solution is intrinsic because the activity of ethanol oxidation depends on the pH [26]. According to stability of ethanol in alkaline solution, we suggested the deprotonation of ethanol in alkaline solution before oxidation reaction on Au surface. In this section, we focused on the ethanol oxidation mechanism on Au catalyst in alkaline solution as a first step to tackle the electro-oxidation of biofuels in fuel cells [54].

4.4.2 Reaction of Ethanol over Au Catalyst

We discuss the reaction mechanism of ethanol oxidation over Au catalyst assuming different surface states: bare, OH-covered, and O-covered Au surfaces. We first analyzed the initial reaction step, i.e., ethanol oxidation to acetaldehyde, over bare Au surface. Because the majority of the ethanol exists as deprotonated $CH_3CH_2O^-$, first reaction step is the following:

$$CH_3CH_2O^- + OH^- \rightarrow CH_3CHO + H_2O + 2e^- \qquad (4.3)$$

In the analysis, we assumed the Eley-Rideal mechanism, i.e. adsorbed $CH_3CH_2O^-$ reacts with OH^- in liquid phase. Figure 4.5 shows the reaction pathway over bare Au surface calculated by using DFT method. $CH_3CH_2O^-$ adsorbs on Au surface via O atom. When OH^- in liquid phase interacts with adsorbed $CH_3CH_2O^-$, the system is stabilized by 54.4 kJ/mol. The activation energy of the subsequent reaction step to form acetaldehyde was calculated to be zero. This indicates that the $CH_3CH_2O^-$ forms acetaldehyde without barrier when it adsorbs and interacts with liquid phase OH^- on Au catalyst surface.

We then analyzed the reaction mechanism of ethanol oxidation over OH^- adsorbed Au surface. Figure 4.6 shows the reaction

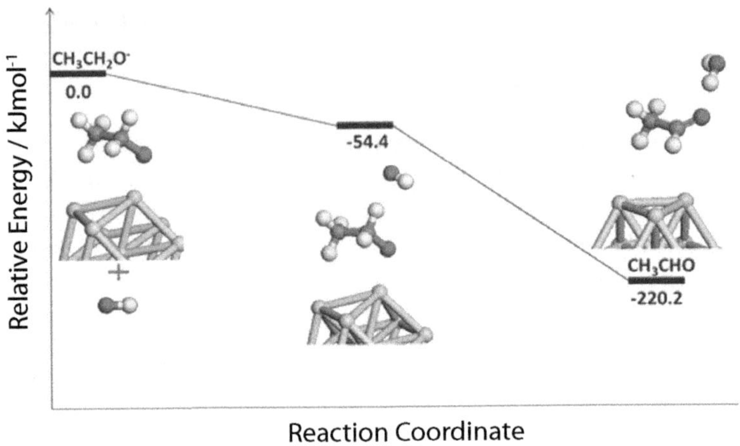

Figure 4.5 Initial reaction step of ethanol oxidation on bare Au surface.

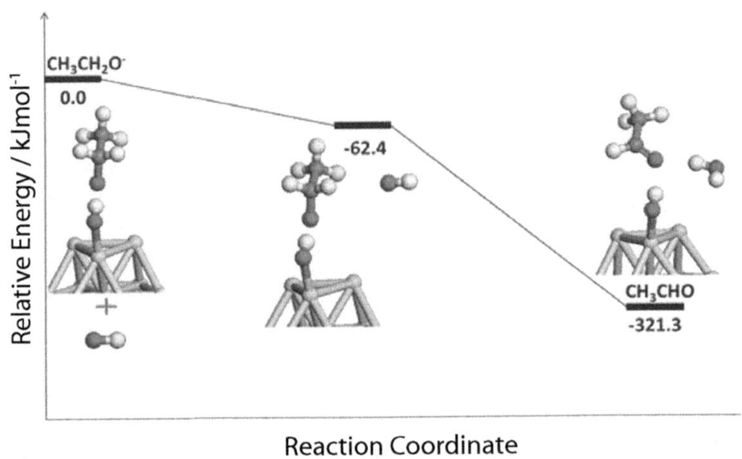

Figure 4.6 Initial reaction step of ethanol oxidation on OH-covered Au surface.

pathway calculated by using DFT method. O of $CH_3CH_2O^-$ interacts with H of OH^- adsorbed on Au surface. Adsorbed $CH_3CH_2O^-$ then interacts with OH^- in liquid phase with the interaction energy of −62.4 kJ/mol. Subsequently the interacting OH^- abstract H from

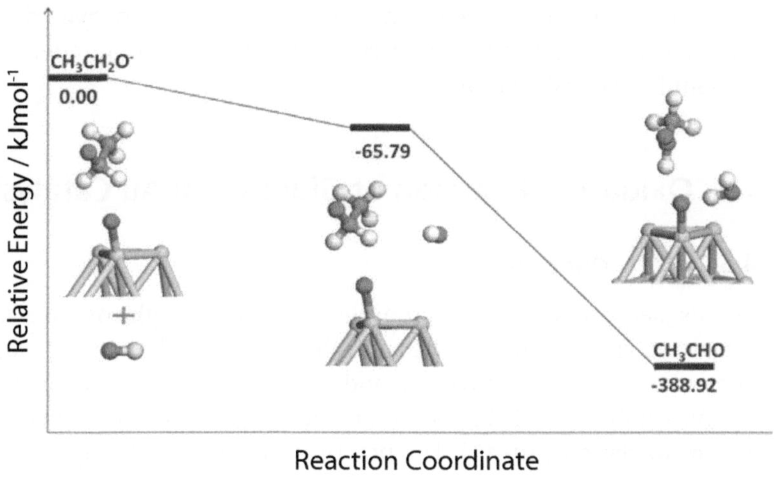

Figure 4.7 Initial reaction step of ethanol oxidation on O-covered Au surface.

$CH_3CH_2O^-$ to form acetaldehyde. No reaction barrier was obtained for this step also on OH^- adsorbed Au surface.

We finally analyzed the reaction mechanism of ethanol oxidation over O-adsorbed Au surface. Figure 4.7 shows the reaction pathway calculated by using DFT method. O on Au interacts with α-H of $CH_3CH_2O^-$. Interaction energy of adsorbed $CH_3CH_2O^-$ and liquid phase OH^- was calculated to be -65.8 kJ/mol. Subsequent H abstraction from $CH_3CH_2O^-$ by OH^- to form acetaldehyde was calculated to have no reaction barrier. We can conclude that $CH_3CH_2O^-$ on Au surface reacts with liquid phase OH^- without barrier to form acetaldehyde regardless of the surface states of Au catalyst.

4.4.3 Summary

We analyzed the ethanol oxidation reaction mechanism on Au catalyst in alkaline solution. In the oxidation reaction of ethanol, we considered three influences of alkaline environment on the reaction, i.e., influences of (i) OH^- in the solution on the ethanol dissociation in liquid phase, (ii) OH^- on surface state of Au catalyst, and (iii) OH^- as an oxidative species in the ethanol oxidation. We clarified that the reaction step from deprotonated ethanol to acetaldehyde does not have significant reaction barrier in alkaline solution under the present level of calculation.

4.5 Oxidation Reaction of Glucose on Au Catalyst

4.5.1 Introduction

In this section, we focused on the oxidation of glucose on Au catalyst in alkaline solution because the glucose is abundant, cheap, nontoxic, and easy to produce and handle. It is well known that the Au catalyst in alkaline solution shows high activity of glucose electro-oxidation [30–41]. For practical operation, although there are many experimental studies for Au nanoparticle sizes and supported materials, the oxidation reaction mechanism of glucose is unclear even the first step of oxidation from glucose to gluconic acid.

$$C_6H_{12}O_6 + 2OH^- \rightarrow C_6H_{12}O_7 + H_2O + 2e^- \quad (4.4)$$

Understanding of glucose oxidation on Au surface in alkaline solution is essential for high active and efficient catalytic material design. Some experimental works suggested the mechanism of glucose electrochemical oxidation on Au surface [58–60]. Among previous works, Larew et al. proposed that the oxidation of aldehyde (CHO) group in glucose proceeds on the OH covered Au surface with OH^- molecule in alkaline solution [59]. Theoretical proof is necessary to confirm the glucose oxidation reaction mechanism. However, to the best of our knowledge, there are no theoretical reports for glucose oxidation reaction mechanism. We analyzed the glucose oxidation reaction mechanism on Au surface in alkaline solution by using DFT method [55]. We also estimated the effect of adsorbate species on Au surface in alkaline solution.

4.5.2 Reaction of Glucose

We analyzed the oxidation reaction mechanism of glucose on in alkaline solution by using Au_{55}-OH cluster model. The open

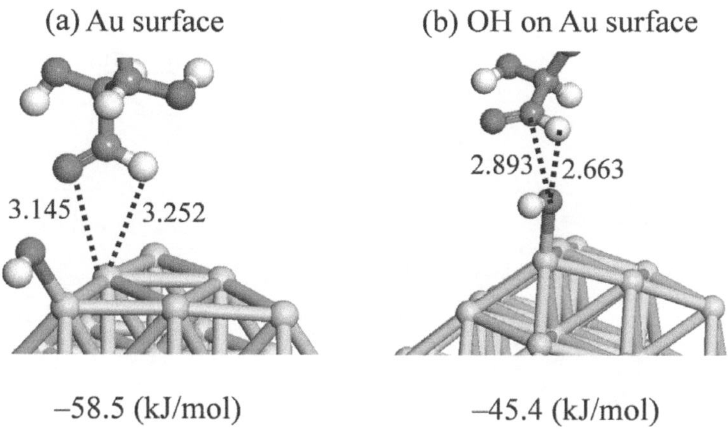

Figure 4.8 Adsorption energy and structure of glucose on Au_{55}-OH cluster.

chain form of glucose was used in this study. The CHO group of glucose is converted to carboxylic acid (COOH) group by the initial oxidation reaction. We considered the adsorption of CHO group, which is oxidized part of glucose, on Au_{55}-OH. Two stable adsorbed structures of glucose on Au_{55}-OH are shown in Fig. 4.8.

One is the direct adsorption of glucose on Au surface (Fig. 4.8a), the other is the adsorption of glucose on OH of Au surface (Fig. 4.8b). The adsorption energy of glucose on Au surface was more stable than that on OH of Au surface by about 13 kJ/mol. The adsorbed structure in Fig. 4.8a is close to the interaction structure of acetaldehyde with Au reported in Ref. [57], which is intermediate structure of oxidation from ethanol to acetic acid in alkaline solution. On the other hand, the adsorbed structure in Fig. 4.8b is a model of interaction structure, which is proposed by the experiment [58]. According to these adsorbed structures of glucose, we analyzed the oxidation reaction pathways from glucose to gluconic acid.

We analyzed the oxidation reaction mechanism assuming direct adsorption of glucose on Au surface (in Fig. 4.8a). The calculated reaction energy diagram is shown in Fig. 4.9.

As a reaction pathway from glucose to gluconic acid on Au surface, we assumed two pathways. One is the H transfer from CHO group to Au surface followed by OH transfer from Au surface

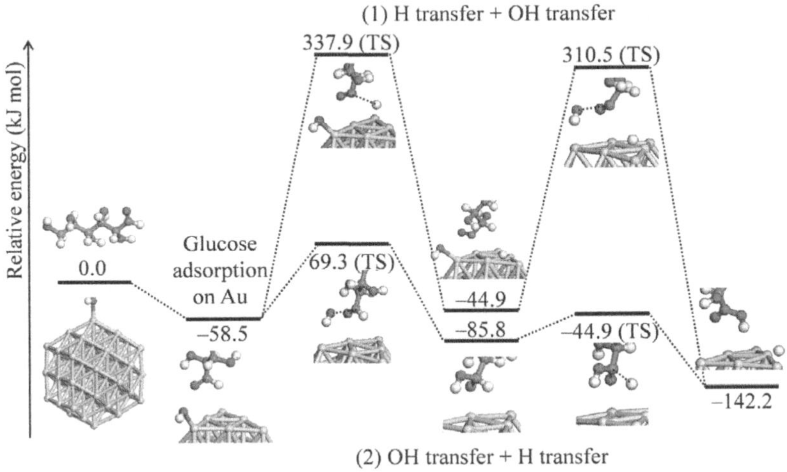

Figure 4.9 Reaction energy diagram of oxidation reaction assuming direct adsorption of glucose on Au surface (Fig. 4.8a).

to CO group in glucose (reaction 1), the other is the OH transfer from Au surface to CHO group in glucose followed by H transfer from CHOOH to Au surface (reaction 2). The first transition state energy in reaction (1) was 396.4 kJ/mol. The next transition state energy for OH transfer from Au to CO group in glucose was also high (355.4 kJ/mol). On the other hand, the first transition state energy for OH transfer from Au to CHO group in glucose in reaction 2 was 127.8 kJ/mol. The gluconic acid was formed through lower energy transition state (40.9 kJ/mol) for H transfer from CHOOH to Au. Four-coordinated carbon was formed in the intermediate structure between two transition state structures in reaction 2. Comparing two reaction pathways, reaction 2 is preferable pathway. The first transition state for OH transfer from Au to CHO group in glucose in reaction 2 of glucose oxidation reaction is rate-determining step.

We next analyzed the oxidation reaction mechanism of adsorbed structure of glucose on OH of Au surface (in Fig. 4.8b). The calculated reaction energy diagram is shown in Fig. 4.10. In this mechanism, we explicitly considered OH^- in alkaline solution as reactive species according to the reaction model proposed by experimental result [58]. The hydrogen bond between CHO group of glucose on Au_{55}-

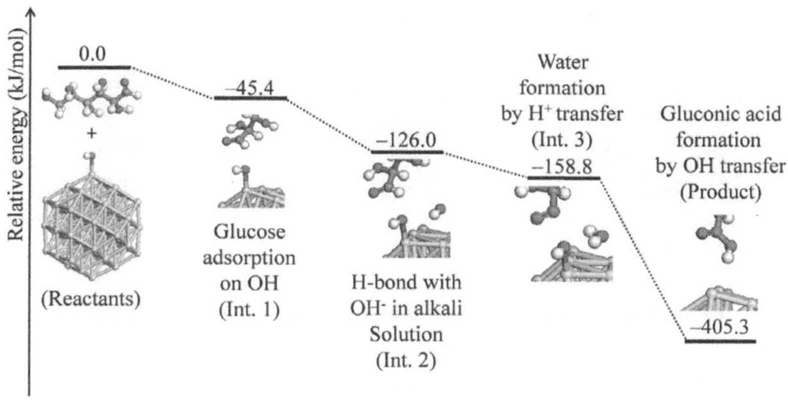

Figure 4.10 Reaction energy diagram of oxidation reaction assuming adsorption of glucose on OH of Au surface (Fig. 4.8b).

OH and OH^- in alkaline solution was formed (Int. 2 in Fig. 4.10). The water molecule was formed by the proton transfer from CHO group of glucose to OH^- in alkaline solution (Int. 3 in Fig. 4.10). The gluconic acid was formed without any transition state in this calculation level of theory (product in Fig. 4.10). From above theoretical analysis, we found the oxidation reaction mechanism of glucose on Au catalyst in alkaline solution theoretically as follows: (i) Glucose adsorbs on OH of Au surface. (ii) OH^- in alkaline solution interacts with CHO group of glucose. (iii) Water is formed by proton transfer from CHO group of glucose to OH^- in solution. (iv) Gluconic acid is formed by OH transfer from Au surface. These reaction proceeds on the OH covered Au catalyst surface in alkaline solution.

4.5.3 Effect of Adsorbate Species on Au Surface

It is important to rationalize the effect for reactivity of glucose oxidation with the different coverage condition by OH^- on Au surface to find the high active Au surface condition. We analyzed the reactivity of oxidized OH covered Au surface and high coverage OH on Au surface by using Au_{13} cluster, which is a model for screening of reaction activity under the reasonable computational costs. The Au_{13} cluster models (Au_{13}-OH, Au_{13}-O, and Au_{13}-$(OH)_3$) are shown in Fig. 4.11.

Figure 4.11 Au_{13} cluster models with different adsorbed species.

Table 4.4 Relative energy change during glucose oxidation reaction on Au_{55}-OH, Au_{13}-OH, Au_{13}-O and Au_{13}-$(OH)_3$ clusters. Unit is in kJ/mol. The reactants Int. 1, Int. 2, Int. 3 and product correspond to the structures in Fig. 4.10

Adsobates	Reactants	Int. 1	Int. 2	Int. 3	Products
Au_{55}-OH	0.0	−45.4	−126.0	−158.8	−405.3
Au_{13}-OH	0.0	−47.3	−130.3	−141.2	−365.5
Au_{13}-O	0.0	−47.8	−121.0	−128.9	−572.6*
Au_{13}-$(OH)_3$	0.0	−53.5	−124.0	−144.6	−288.3

*Gluconic acid anion ($C_6H_{11}O_7^-$) was formed as the final product.

Au_{13}-O is an oxidized model structure by proton transfer from OH on Au surface to OH^- in alkaline solution. Au_{13}-$(OH)_3$ is a model structure of high OH^- coverage condition on Au surface. We calculated the oxidation reaction of glucose on these Au_{13} cluster models based on the reaction energy diagram in Fig. 4.10. The stable energies (corresponding to reactants, Int. 1, Int. 2, Int. 3, and Product in Fig. 4.10) of each Au_{13} cluster model are listed in Table 4.4. Although the energy difference of final product between Au_{55}-OH and Au_{13}-OH was about 40 kJ/mol, the energy difference of each step was relatively small. The result obtained by Au_{13}-OH model is reasonable to estimate the effect of reactivity of glucose oxidation.

We next analyzed the reaction energy change of glucose oxidation on Au_{13}-O. The adsorption energies (−47.3 and −47.8 kJ/mol) of glucose on Au_{13}-OH and Au_{13}-O were almost same.

The interaction energy between CHO group of glucose and OH^- in alkaline solution on Au_{13}-O was about 10 kJ/mol smaller than that on Au_{13}-OH. Also, the intermediate structure for water formation by proton transfer on Au_{13}-O was about 20 kJ/mol smaller than that on Au_{13}-OH.

This result indicates that the Au_{13}-OH shows high glucose oxidation reaction activity compared to Au_{13}-O due to the difference of energy stability of intermediate structures. Here, Assiongbon et al. experimentally found that the oxidation activity on Au-O is lower than Au-OH surface [55]. Our result is good agreement with experimental trend of surface oxidation. In the case of Au_{13}-$(OH)_3$, the adsorption energy of glucose (−53.5 kJ/mol) was larger than that on Au_{13}-OH. The energy difference for proton transfer on Au_{13}-$(OH)_3$ (−20.6 kJ/mol) was larger than that on Au_{13}-OH (−10.9 kJ/mol). We found that the oxidized OH covered Au surface is poor for glucose oxidation. The high active catalytic property of Au for glucose oxidation in alkaline solution is important to keep the high coverage OH on Au surface.

4.5.4 Summary

We analyzed the glucose oxidation reaction mechanism from different adsorbed structures between glucose and Au_{55}-OH. The adsorption of glucose on Au surface was more stable than that on OH of Au surface. The barrier in reaction from direct adsorption of glucose on Au surface was relatively high, and we found that the first transition state for OH transfer from Au to CHO group of glucose is rate-determining step. On the other hand, there are no obvious transition states in the reaction from glucose adsorption on OH of Au. We showed that the glucose oxidation reaction will occur spontaneously on Au-OH surface to consider direct interaction and reaction of glucose and OH^- in alkaline solution. In addition, we evaluated the effect of Au surface structure for glucose oxidation reaction. Schematic catalytic reaction cycle for glucose is shown in Fig. 4.12. We found that the high OH covered Au surface is favorable of glucose oxidation.

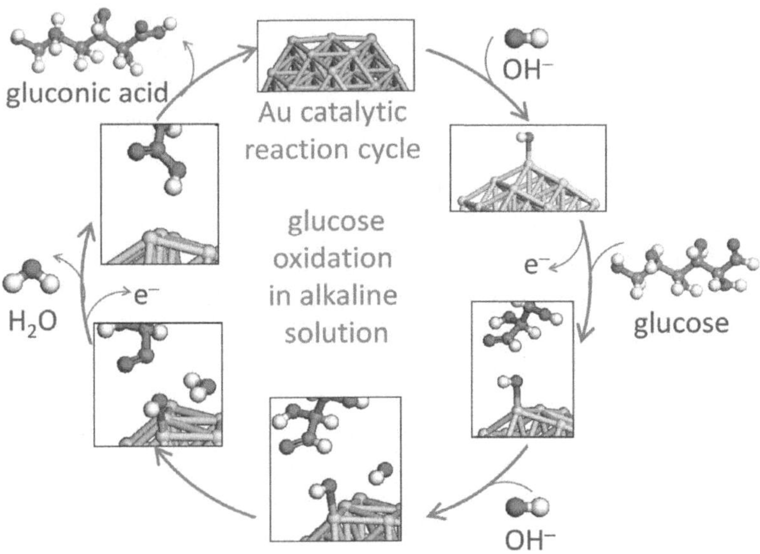

Figure 4.12 Schematic illustration of Au catalytic cycle for glucose oxidation reaction in alkaline solution.

4.6 Effect of Support Materials for Au Catalyst

4.6.1 Introduction

In addition to intrinsic catalytic activity of Au, the choice of support materials for Au catalyst is important to improve the catalytic activity [61–72]. The carbon based materials such as glassy carbon, highly ordered pyrolytic graphite, carbon nanotubes, and graphene has attracted interest in many applications such as electrocatalysis, molecular electronics, biosensing, and optical devices [61–67]. On the other hand, the various oxides have been recognized as the effective support material for catalysis [68–72]. A wide range of supports have been tested, such as TiO_2, ZrO_2, SnO_2, and CeO_2. In fact, CO oxidation on Au catalyst was improved by SnO_2 support [72]. One will often observe a certain support facilitates specific reactions while it retards other reactions. We analyzed the interaction energy and charge transfer between support materials, such as carbon and oxides, and Au catalyst by using DFT method [73].

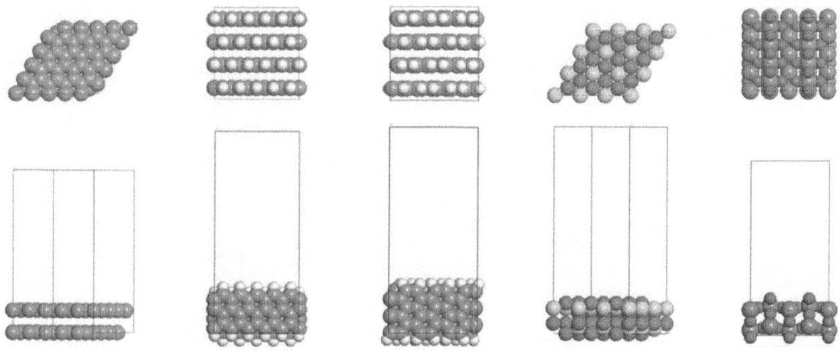

Figure 4.13 Model structures of carbon (graphite(0001), (10$\bar{1}$0), and (11$\bar{2}$0)) and oxide (ZrO_2(111) and (SnO_2(110)) surfaces as support materials.

4.6.2 Interaction between Support Materials and Au Catalyst

We first analyzed the interaction between Au_{13} cluster and the support materials. We show in Fig. 4.13 the slab models of carbon/oxide support materials. In this study, basal and edge (zigzag and armchair) planes, which correspond to graphite(0001), (10$\bar{1}$0), and (11$\bar{2}$0) surfaces, are used as carbon support model. The dangling bonds of (10$\bar{1}$0) and (11$\bar{2}$0) surfaces were terminated by hydrogen. We used the ZrO_2(111) and SnO_2(110) surfaces as the oxide support model for Au_{13} cluster.

The optimized structures of Au_{13} cluster on the support materials are shown in Fig. 4.14. Significant structure changes of Au_{13} clusters were not observed thus the Au_{13} clusters on support materials keep the spherical shape likewise the optimized structure of isolated Au_{13} cluster. The interaction energy of Au_{13} cluster with carbon and oxide surfaces is shown in Table 4.5.

All interaction energies between Au_{13} cluster and support surfaces were negative indicating the stabilization of Au_{13} cluster on the selected support materials. The graphite(11$\bar{2}$0) surface shows the most stable interaction energy of Au_{13} cluster in the carbon surfaces studied here. To find the difference of interaction energy, we analyzed the charge transfer between Au_{13} cluster and carbon/oxide

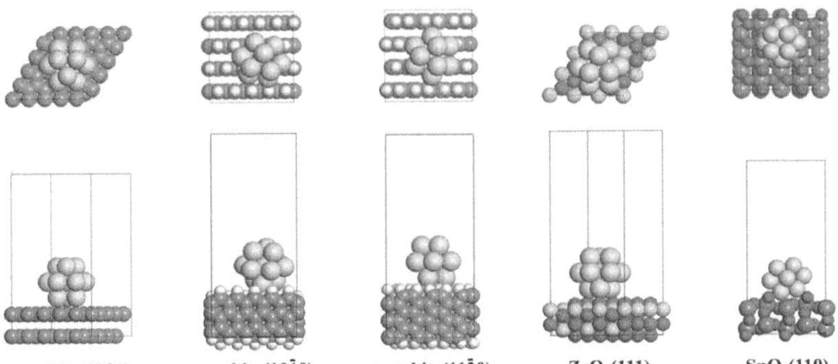

Figure 4.14 Optimized structures of Au_{13} clusters on carbon (graphite(0001), (10$\bar{1}$0), and (11$\bar{2}$0)) and oxide (ZrO_2(111) and (SnO_2(110)) surfaces.

Table 4.5 Interaction energy (kJ/mol) between Au_{13} cluster and carbon (graphite (0001), (10$\bar{1}$0), and (11$\bar{2}$0)) and oxide (ZrO_2(111) and (SnO_2(110)) surfaces

Surface	Interaction energy (kJ/mol)
graphite(0001)	−151.5
graphite(10$\bar{1}$0)	−341.6
graphite(11$\bar{2}$0)	−581.0
ZrO_2(111)	−157.3
SnO_2(110)	−396.7

support materials. The atomic charges of support materials and supported Au_{13} cluster are shown in Table 4.6.

The atomic charges of support materials and Au_{13} cluster were neutral before interaction. In the case of carbon surfaces, the electrons in graphite(10$\bar{1}$0) and (11$\bar{2}$0) surfaces transferred to Au_{13} clusters. Due to the smaller electron transfer between Au_{13} cluster and graphite(0001) surface, carbon(0001) surface has smaller interaction energy. Contrary to the carbon and ZrO_2(111) surfaces, the electrons in Au_{13} clusters transferred to SnO_2(110) surface.

Table 4.6 Atomic charge of support materials (carbon (graphite(0001), $(10\bar{1}0)$, and $(11\bar{2}0)$) and oxide (ZrO_2(111) and (SnO_2(110))) and Au_{13} cluster after interaction

Surface	Support material	Au_{13} cluster
graphite(0001)	0.07	−0.07
graphite$(10\bar{1}0)$	0.62	−0.62
graphite$(11\bar{2}0)$	0.91	−0.91
ZrO_2(111)	0.46	−0.46
SnO_2(110)	−1.13	1.13

4.6.3 Glucose Oxidation Reaction on Supported Au Catalyst

We analyzed the glucose oxidation reaction pathway on Au catalyst supported by carbon and oxide surfaces. First, we investigated the glucose oxidation reaction on Au catalyst supported by carbon surface. The obtained reaction energy diagram is shown in Fig. 4.15. The stable intermediate structure of each step for graphite(0001) case is also shown in Fig. 4.15 as an example.

The adsorption energy of glucose on OH of graphite(0001) supported Au catalyst was −44.8 kJ/mol. Also, the interaction energy between glucose and OH^- in alkaline solution by hydrogen bond was about −135.1 kJ/mol. Significant difference between graphite(0001) and graphite$(10\bar{1}0)$ and $(11\bar{2}0)$ in these two intermediate states was not observed because the energy difference was within 10 kJ/mol. The support effect of carbon was small up to this reaction step. Concerning the last two steps in the case of graphite$(10\bar{1}0)$ and $(11\bar{2}0)$, obtained relative reaction energies were unstable in comparison with the case of only Au_{13} cluster model with PBC. On the other hand, the relative reaction energies of graphite(0001) were almost same or slightly stable in comparison with the case of only Au_{13} cluster model with PBC. We clearly found the significant difference of reaction energy diagram by the change of surface structure of carbon.

Next, we analyzed the glucose oxidation reaction on Au catalyst supported by oxide surface. The obtained reaction energy diagram is shown in Fig. 4.16. The stable intermediate structure of each step

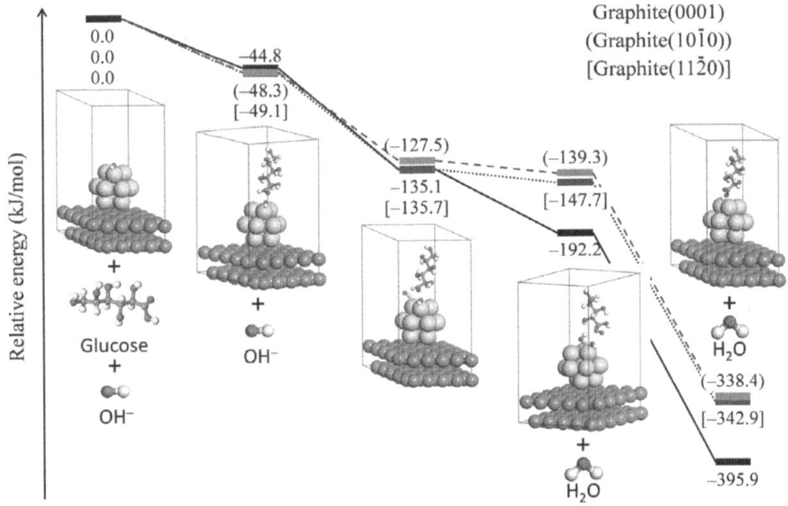

Figure 4.15 Reaction energy diagram of glucose oxidation reaction on carbon (graphite(0001), (10$\bar{1}$0), and (11$\bar{2}$0)) supported Au catalyst. The optimized intermediate structures for graphite(0001) case are also shown.

for $ZrO_2(111)$ case is shown in Fig. 4.16 as an example of stable structure.

As well as the case of graphite, similar relative energy values were obtained for intermediate state of glucose adsorption on Au_{13}-OH and OH^- interaction with glucose. While the water formation energy for $ZrO_2(111)$ was similar to those for graphite(10$\bar{1}$0) and (11$\bar{2}$0) surfaces, the relative energies in the case of $SnO_2(110)$ surface were largely stabilized. As a result, the formation energy of gluconic acid by OH transfer from Au was also largely stabilized in the case of $SnO_2(110)$ surface. According to the linear free-energy relationship for the reactivity estimation, it is assumed that the activity of lower reaction energy path is higher [74]. This result indicates that the $SnO_2(110)$ surface shows high activity as support materials of Au catalyst.

Here, we discuss the reactivity difference for studied support materials of Au catalyst. The carbon and $ZrO_2(111)$ surfaces do not show the improvement of reactivity for glucose oxidation reaction because the reaction energy diagram for these surfaces are similar trend in the case of unsupported Au catalyst. On the other hand, the

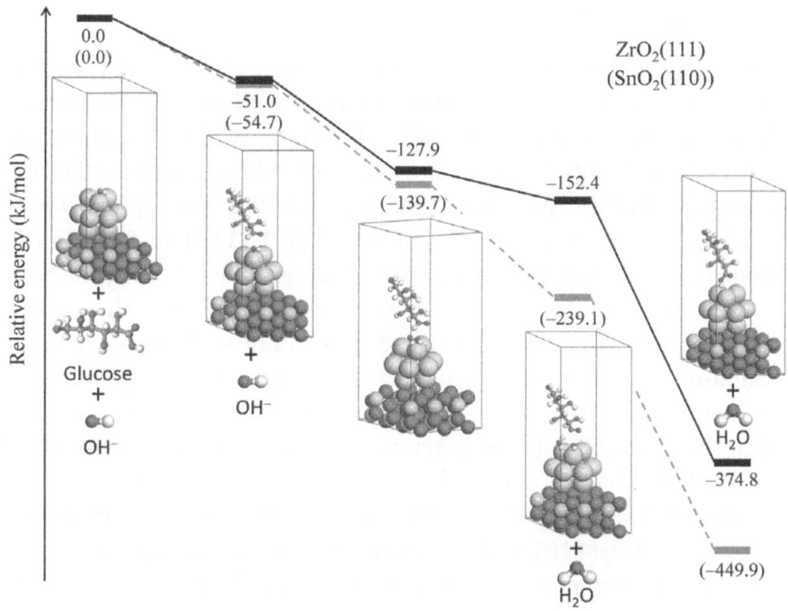

Figure 4.16 Reaction energy diagram of glucose oxidation reaction on oxide ($ZrO_2(111)$ and ($SnO_2(110)$)) supported Au catalyst. The optimized intermediate structures for $ZrO_2(111)$ case are also shown.

$SnO_2(110)$ surface shows large improvement of catalytic reactivity for glucose oxidation reaction because the gluconic acid formation energy and important intermediate structure for water formation by the proton transfer were largely stabilized. One of the main factors of these differences is charge transfer between Au catalyst and support materials. When the electron transfers from support materials to Au catalyst, the reactivity of Au catalyst becomes low. The carbon and $ZrO_2(111)$ surfaces are categorized in this group. When the electron transfers from Au catalyst to support materials, the reactivity of Au catalyst becomes high. The $SnO_2(110)$ surface are categorized in this group. Electron transfer reaction from glucose to Au catalyst is rate-determining process in glucose oxidation reaction. When the Au catalyst has electronically positive charge, the glucose oxidation reaction proceeds smoothly.

4.6.4 Summary

We analyzed the glucose oxidation reaction mechanism and reaction activity of Au catalyst supported by carbon and oxide in alkaline solution environment by using DFT calculation. We first estimated the interaction energy between carbon and oxide supported materials and Au_{13} clusters. We studied the support effect of graphite(0001), $(10\bar{1}0)$, and $(11\bar{2}0)$ and $ZrO_2(111)$ and $SnO_2(110)$ surfaces in this study. Large interaction energy was obtained when the large number of electrons transferred between support materials and Au_{13} cluster. The support effect of carbon and $ZrO_2(111)$ surfaces for glucose oxidation reaction was small because the reaction energy diagram of Au catalyst without support materials was almost same. We found that the glucose oxidation reaction proceeds smoothly by the use of $SnO_2(110)$ surface because the Au_{13} cluster on $SnO_2(110)$ surface is electronically positive due to the electron transfer from Au_{13} cluster to $SnO_2(110)$ surface. To design the high catalytic reactivity by the support materials with metal catalyst, it is important to control the charge transfer between support materials and metal catalyst.

4.7 Summary and Perspectives

In this chapter, we summarized the recent activities on the oxidation reaction of biofuels, such as ethanol and glucose, on Au catalyst in direct alkaline fuel cell from the theoretical point of views. We confirmed three aspects of the effects of alkaline environment on the catalytic process: (i) effect on the states of gold surface, (ii) effect on the states of reactants, and (iii) effect of OH^- species as oxidant. The reaction mechanisms of ethanol and glucose oxidation on Au catalyst in alkaline solution are clearly discussed. By considering the above three effects explicitly, unique catalytic activity of gold in alkaline environment against biofuels is clearly interpreted. As an effort for the improvement of activity, we have introduced the effect of support material on the Au catalysis, which indicates candidate support material for engineering the higher activity.

Acknowledgments

The activities of INAMORI Frontier Research Center are supported by KYOCERA Corporation.

References

1. Ishimoto, T., Ogura, T., Koyama, M., Yang, L., Kinoshita, S., Yamada, T., Tokunaga, M., and Kitagawa, H. (2013). A key mechanism of ethanol electrooxidation reaction in a noble-metal-free metal-organic framework, *J. Phys. Chem. C*, **117**, 10607–10614.
2. Mattos, L. V., Jacobs, G., Davis, B. H., and Noronha, F. B. (2012). Production of Hydrogen from ethanol: review of reaction mechanism and catalyst deactivation, *Chem. Rev.*, **112**, 4094–4123.
3. Aurélien, H., Karine, S., Sophie, T., and Boniface, K. (2011). Performance of enzymatic glucose/O_2 biofuel cells, *Biofuels Eng. Proc. Technol.*, **1**, 467–492.
4. Yang, L., Kinoshita, S., Yamada, T., Kanda, S., Kitagawa, H., Tokunaga, M., Ishimoto, T., Ogura, T., Nagumo, R., Miyamoto, A., and Koyama, M. (2010) A Metal-Organic Framework as an electrocatalyst for ethanol oxidation, *Angew. Chem. Int. Ed.*, **49**, 5348–5351.
5. Nørskov, J. K. Bligaad, T., Rossmeisl, J., and Christensen, C. H. (2009). Towards the computational design of solid catalysts, *Nat. Chem.*, **1**, 37–46.
6. Chaudhuri, S. K. and Lovely, D. R. (2003). Electricity generation by direct oxidation of glucose in mediatorless microbial fuel cells, *Nat. Biotechnol.*, **1**, 1229–1232.
7. Yu, E. H., Wang, X., Krewer, U., Li, L., and Scott, K. (2012). Direct oxidation alkaline fuel cells: from materials to systems, *Energy Environ. Sci.*, **5**, 5668–5680.
8. Zhao, T. S., Li, Y. S., and Shen, S. Y. (2010). Anion-exchange membrane direct ethanol fuel cells: Status and perspective, *Front. Energy Power Eng. China*, **4**, 443–458.
9. Yu, E. H., Krewer, U., and Scott, K. (2010). Principles and materials aspects of direct alkaline alcohol fuel cells, *Energies*, **3**, 1449–1528.
10. Braunchweig, B., Hibbitts, D., Neurock, M., and Wieckowski, A. (2013). Electrocatalysis: a direct alcohol fuel cell and surface science perspective, *Catal. Today*, **202**, 197–209.

11. Friedl, J. and Stimming, U. (2013). Model catalyst studies of hydrogen and ethanol oxidation for fuel cells, *Electrochim. Acta*, **101**, 41–58.
12. Kavanagh, R., Cao, X. M., Lin, W., Hardacre, C., and Hu, P. (2012). Acetaldehyde production in the direct ethanol fuel cell: mechanistic elucidation by density functional theory, *J. Phys. Chem. C*, **116**, 7185–7188.
13. Antolini, E. and Gonzalez, E. R. (2010). Alkaline direct alcohol fuel cells, *J. Power Sources*, **195**, 3431–3450.
14. Lai, S. C. S., Kleijn, S. E. F., Öztürk, F. T. Z., van Rees Vellinga, V. C., Koning, J., Rodriguez, P., and Koper, M. T. M. (2010). Effects of electrolyte pH and composition on the ethanol electro-oxidation reaction, *Catal. Today*, **154**, 92–104.
15. Kowal, A., Li, M., Shao, M., Sasaki, K., Vukmirovic, M. B., Zhang, J., Marinkovic, N. S., Liu, P., Frenkel, A. I., and Adzic, R. R. (2009). Ternary Pt/Rh.SnO_2 electrocatalysts for oxidizing ethanol to CO_2, *Nat. Mater.*, **8**, 325–330.
16. Lai, S. C. S. and Koper, M. T. M. (2009). Ethanol electro-oxidation on platinum in alkaline media, *Phys. Chem. Chem. Phys.*, **11**, 10446–10456.
17. Lai, S. C. S., Kleyn, S. E. F, Rosca, V., and Koper, M. T. M. (2008). Mechanism of the dissociation and electrooxidation of ethanol and acetaldehyde on platinum as studied by SERS, *J. Phys. Chem. C*, **112**, 19080–19087.
18. Simões, F. C., dos Anjos, D. M., Vigier, F., Léger, J. M., Hahn, F., Coutanceau, C., Gonzalez, E. R., Tremiliosi-Filho, G., de Andrade, A. R., Olivi, P., and Kokoh, K. B. (2007). Electroactivity of tin modified platinum electrodes for ethanol electrooxidation, *J. Power Sources*, **167**, 1–10.
19. Durbin, D. J. D. and Malardier-Jugroot, C. (2011). Density functional theory analysis of metal/graphene systems as a filter membrane to prevent CO poisoning in hdrogen fuel cells, *J. Phys. Chem. C*, **115**, 808–815.
20. Ponrouch, A., Garbarino, S., and Guay, D. (2009). Effect of the nanostructure on the CO poisoning rate of platinum, *Electrochem. Commun.*, **11**, 834–837.
21. Das, S. K., Reis, A., and Berry, K. J. (2009). Experimental evaluation of CO poisoning on the performance of a high temperature proton exchange membrane fuel cell, *J. Power Sources*, **193**, 691–698.
22. Koper, M. T. M. (2013). A basic solution, *Nat. Chem.*, **5**, 255–256.
23. Ide, M. S. and Davis, R. J. (2014). The important role of hydroxyl on oxidation catalysis by gold nanoparticles, *Acc. Chem. Res.*, **47**, 825–833.

24. Zhang, J., Liang, Y., Li, N., Li, Z., Xu, C., and Jiang, S. P. (2012). A remarkable activity of glycerol electrooxidation on gold in alkaline medium, *Electrochim. Acta*, **59**, 156–159.

25. Zhang, Z., Xin, L., and Li, W. (2012). Supported gold nanoparticles as anode catalyst for anion-exchange membrane-direct glycerol fuel cell (AEM-DGFC), *Int. J. Hydrogen Energy*, **37**, 9393–9401.

26. Rodriguez, P., Kwon, Y., and Koper, M. T. M. (2012). The promoting effect of adsorbed carbon monoxide on the oxidation of alcohols on a gold catalyst, *Nat. Chem.*, **4**, 177–182.

27. Kwon, Y., Lai, S. C. S., Rodriguez, P., and Koper, M. T. M. (2011). Electrocatalytic oxidation of alcohols on gold in alkaline media: base or gold catalysis?, *J. Am. Chem. Soc.*, **133**, 6914–6917.

28. Thompson, D. T. (2007). Using gold nanoparticles for catalysis, *Nano Today*, **2**, 40–43.

29. Li, L., Scott, K., and Yu, E. H. (2013). A direct glucose alkaline fuel cell using MnO_2-carbon nanocomposite supported gold catalyst for anode glucose oxidation, *J. Power Sources*, **221**, 1–5.

30. Tung, A., Hunag, T., Lee, C., and Chiu, H. (2012). Electrochemical growth of gold nanostructures on carbon paper for alkaline direct glucose fuel cell, *RSC Adv.*, **2**, 1068–1073.

31. Cuevas-Muñiz, F. M., Guerra-Balcázar., Castaneda, F., Ledesma-García, L., and Arriaga, L. G. (2011). Performance of Au and AuAg nanoparticles supported on Vulcan in a glucose laminar membraneless microfuel cell, *J. Power Sources*, **196**, 5853–5857.

32. An, L., Zhao, T. S., Shen, S. Y., Wu, Q. X., and Chen, R. (2011). Alkaline direct oxidation fuel cell with non-platinum catalysts capable of converting glucose to electricity at high power output, *J. Power Sources*, **196**, 186–190.

33. Habrioux, A., Servat, K., Girardeau, T., Guérin, P., Napporn, T. W., and Boniface Kokoh, K. (2011). Activity of sputtered gold particles layers towards glucose electrochemical oxidation in alkaline medium, *Curr. Appl. Phys.*, **11**, 1149–1152.

34. Basu, D. and Basu, S. (2010). A study on direct glucose and fructose alkaline fuel cell, *Electrochim. Acta*, **55**, 5775–5779.

35. Pasta, M., Ruffo, R., Falletta, E., Mari, C. M., and Pina, C. D. (2010). Alkaline glucose oxidation on nanostructured gold electrodes, *Gold Bull.*, **43**, 57–64.

36. Liu, Z., Huang, L., Zhang, L., Ma, H., and Ding, Y. (2009). Electrocatalytic oxidation of D-glucose at nanoporous Au and Au-Ag alloy electrodes in alkaline aqueous solutions, *Electrochim. Acta*, **54**, 7286–7293.

37. Ishida, T., Kinoshita, N., Okatsu, H., Akita, T., Takei, T., and Haruta, M. (2008). Influence of the support and the size of gold clusters on catalytic activity for glucose oxidation, *Angew. Chem. Int. Ed.*, **47**, 9265–9268.
38. Tominaga, M., Shimazoe, T., Nagashima, M., and Taniguchi, I. (2005). Electrocatalytic oxidation of glucose at gold nanoparticle-modified carbon electrodes in alkaline and neutral solutions, *Electrochem. Commun.*, **7**, 189–193.
39. Önal, Y., Schimpf, S., and Claus, P. (2004). Structure sensitivity and kinetics of D-glucose oxidation to D-gluconic acid over carbon-supported gold catalysts, *J. Catal.*, **223**, 122–133.
40. Martins, A., Ferreira, V., Queirós, A, Aroso, I., Silva, F., and Fuliu, J. (2003). Enantiomeric electro-oxidation of D-and L-glucose on chiral gold single crystal surfaces, *Electrochem. Commun.*, **5**, 741–746.
41. Biella, S., Prati, L., and Rossi, M. (2002). Selective oxidation of D-glucose on gold catalyst, *J. Catal.*, **206**, 242–247.
42. Perdew, J. P., Chevary, J. A., Vosko, S. H., Jackson, K. A., Pederson, M. R., Singh, D. J., and Fiolhais, C. (1993). Erratum: atoms, molecules, solids, and surfaces: applications of the generalized gradient approximation for exchange and correlation [*Phys. Rev. B*, **46**, 6671 (1992)], *Phys. Rev. B*, **48**, 4978.
43. Perdew, J. P., Chevary, J. A., Vosko, S. H., Jackson, K. A., Pederson, M. R., Singh, D. J., and Fiolhais, C. (1992). Atoms, molecules, solids, and surfaces: applications of the generalized gradient approximation for exchange and correlation, *Phys. Rev. B*, **46**, 6671–6687.
44. Delley, B (2000). From molecules to solids with the DMol^3 approach, *J. Chem. Phys.*, **113**, 7756–7764.
45. Delley, B. (1990). An all-electron numerical method for solving the local density functional for polyatomic molecules, *J. Chem. Phys.*, **92**, 508–517.
46. Klamt, A. and Schuurmann, G. (1993). COSMO: a new approach to dielectric screening in solvents with explicit expressions for the screening energy and its gradient, *J. Chem. Soc., Perkin Trans.*, **2**, 799–805.
47. Govind, N., Pererson, M., Fitzgerald, G., King-Smith, D., and Andzelm, J. (2003). A generalized synchronous transit method for transition state location, *Comp. Meter. Sci.*, **28**, 250–258.
48. Halgren, T. A. and Lipscomb, W. N. (1977). The symchronous-transit method for determining reaction pathways and locating molecular transition states, *Chem. Phys. Lett.*, **49**, 225–232.

49. Hirshfeld, F. L. (1977). Bonded-atom fragments for describing molecular charge densities, *Theoret. Chim. Acta.*, **44**, 129–138.
50. Pessoa, A. M., Fajín, J. L. C., Gomes, J. R. B., and Cordeiro, M. N. D. S. (2012). Ionic and radical adsorption on the Au(*hkl*) surfaces: a DFT study, *Surf. Sci.*, **606**, 69–77.
51. Boronat, M., Corma, A., Illas, F., Radilla, J., Ródenas, T., and Sabater, M. J. (2011). Mechanism of selective alcohol oxidation to aldehyde on gold catalysts: Influence of surface roughness on reactivity, *J. Catal.*, **278**, 50–58.
52. Passoa, A. M., Fajín, J. L. C., Gomes, J. R. B., and Cordeiro, M. N. D. S. (2010). Cluster and periodic DFT calculations of adsorption of hydroxyl on the Au(*hkl*) surfaces, *J. Mol. Struct. (THEOCHEM)*, **946**, 43–50.
53. Häberien, O. D., Chung, S., Stener, M., and Rösch, N. (1997). From clusters to bulk: A relativistic density functional investigation on a series of gold clusters Au_n, $n = 6, \ldots, 147$, *J. Chem. Phys.*, **106**, 5189–5201.
54. Koyama, M., Amano, Y., and Ishimoto, T. (2012). Reaction mechanism of ethanol oxidation over gold catalyst under alkaline solution, *ECS Trans.*, **50**(2), 1907–1912.
55. Ishimoto, T., Kazuno, H., Kishida, T., and Koyama, M. (2014). Theoretical study on oxidation reaction mechanism on Au catalyst in direct alkaline fuel cell, *Solid State Ionics*, **262**, 328–331.
56. Assiongbon, K. A. and Roy, D. (2005). Electro-oxidation of methanol on gold in alkaline media: adsorption characteristics of reaction intermediates studied using time resolved electro-chemical impedance and surface plasmon resonance techniques, *Surf. Sci.*, **594**, 99–119.
57. Zope, B. N., Hibbitts D. D., Neurock, M., and davis, R. J. (2010). Reactivity of the gold/water interface during selective oxidation catalysis, *Science*, **330**, 74–77.
58. Pasta, M., Mantia, F. L., and Cui, Y. (2010). Mechanism of glucose electrochemical oxidation on gold surface, *Electrochim. Acta*, **55**, 5561–5568.
59. Larew, L. A. and Johnson, D. C. (1989). Concentration dependence of the mechanism of glucose oxidation at gold electrodes in alkaline media, *J. Electroanal. Chem*, **262**, 167–182.
60. Makovos, E. B. and Liu, C. C. (1986). 843-A cyclic-voltammetric study of glucose oxidation on a gold electrode, *Biochem. Bioenerg.*, **15**, 157–165.
61. Nakamura, M., Irie, M., Yuge, R., Ichihashi, T., Iijima, S., and Yudasaka, M. (2013). Carboxylation of thin graphitic sheets is faster than that of carbon nanohorns, *Phys. Chem. Chem. Phys.*, **15**, 16672–16675.

62. Kocak, I., Ghanem, M. A., Al-Mayouf, A., Alhoshan, M., and Bartlett, P. N. (2013). A study of the modification of glassy carbon and edge and basal plane highly oriented pyrolytic graphite electrodes modified with anthraquinone using diazonium coupling and solid phase synthesis and their use for oxygen reduction, *J. Electroanal. Chem.*, **706**, 25–32.

63. Stevanović, A. I., Panić, V. V., Dekanski, A. B., Tripković, A. V., and Jovanović, V. M. (2012). Relationships between structure and activity of carbon as a multifunctional support for electrocatalysts, *Phys. Chem. Chem. Phys.*, **14**, 9475–9485.

64. Fu, W., Kiggans, J., Overbury, S. H., Schwartz, V., and Liang, C. (2011). Low-temperature exfoliation of multilayer-graphene material from $FeCl_3$ and CH_3NO_2 co-intercalated graphite compound, *Chem. Commun.*, **47**, 5265–5267.

65. Yan, S., Zhang, S., and Liu, G. (2011). Electrocatalytic performance of gold nanoparticles supported on activated carbon for methanol oxidation in alkaline solution, *J. Phys. Chem. C*, **115**, 6986–6993.

66. Schwartz, V., Xie, H., Meyer III, H. M., Overbury, S. H., and Liang, C. (2011). Oxidative dehydrogenation of isobutene on phosphorous-modified graphitic mesoporous carbon, *Carbon*, **49**, 659–668.

67. Qin, Y., Yang, H., Zhang, X., Li, P., and Ma, C. (2010). Effect of carbon nanofibers microstructure on electrocatalytic activities of Pd electrocatalysts for ethanol oxidation in alkaline medium, *Int. J. Hydrogen Energy*, **35**, 1667–1674.

68. Kim, H. Y., Lee, H. M., and Henkelman, G. (2012). CO oxidation mechanism on CeO_2-supported Au nanoparticles, *J. Am. Chem. Soc.*, **134**, 1560–1570.

69. Jiang, D., Overbury, S. H., and Dai, S. (2011). Interaction of gold clusters with a hydroxylated surface, *J. Phys. Chem. Lett.*, **2**, 1211–1215.

70. Hammer, N., Mathisen, K., Zscherpe, T., Chen, D., and Rønning, M. (2011). Effect of pretreatment on carbon-supported Au/TiO_2 catalysts for preferential oxidation of CO, *Top. Catal.*, **54**, 922–930.

71. Galeano, C., Guttel, R., Paul, M., Aenal, P., Lu, A., and Schüth, F. (2011). Yolk-shell gold nanoparticles as model materials for support-effect studies in heterogeneous catalysis: Au, @C and Au, @ZrO_2 for CO oxidation as an example, *Chem. Eur. J.*, **17**, 8434–8439.

72. Ye, Q., Wang, J., Zhao, J., Yan, L., Cheng, S., Kang, T., and Dai, H. (2010). Pt or Pd-doped Au/SnO_2 catalysts: high activity for low temperature CO oxidation, *Catal. Lett.*, **138**, 56–61.

73. Ishimoto, T., Hamatake, Y., Kazuno, H., Kishida, T., and Koyama, M. (2015). Theoretical study on support effect of Au catalyst for glucose oxidation of alkaline fuel cell anode, *Appl. Surf. Sci.*, **324**, 76–81.

74. Bligaard, T., Nørskov, J. K., Dahl, S., Chrinstensen, C. H., and Sehested, J. (2004). The Bronsted-Evans-Polanyi relation and the volcano curve in heterogeneous catalysis, *J. Catal.*, **224**, 206–217.

Chapter 5

Proton Transport and Design of Proton Electrolyte Membranes for Direct Alcohol Fuel Cells

Liuming Yan and Baohua Yue
Department of Chemistry, College of Sciences, Shanghai University, 99 Shangda Road, Shanghai 200444, China
liuming.yan@shu.edu.cn

This chapter covers the topic of proton, alcohol, and water transport mechanism, and the rational design of proton electrolyte membranes (PEMs). In the first section, we give a brief background description of PEMs. And then, we discuss the types of PEMs. In the third section, the proton, alcohol, and water transport mechanisms is discussed and design procedure of PEMs is proposed. In the fourth section, we give an example of rational design of PEMs, including the study of transport characteristics, the screen and choice of functional groups, and the proof-of-concept experiments. In the last section, we summarize the major points covered in the chapter.

Nanomaterials for Direct Alcohol Fuel Cell
Edited by Yixuan Wang

ISBN 978-981-4669-00-9 (Hardcover), 978-981-4669-01-6 (eBook)
www.panstanford.com

5.1 Introduction

Direct alcohol fuel cells (DAFCs) are attractive alternatives to the proton electrolyte membrane fuel cells (PEMFCs) and rechargeable batteries, especially lithium ion batteries for portable electric and electronic devices, owing to their structural compactness, low-operational temperature, high energy and power density, and readiness of alcohol supply chain. In the core of any DAFCs, there is the membrane electrode assembly (MEA) consisting of a polymeric electrolyte membrane in the middle, an anodic electrocatalyst layer, and a cathodic electrocatalyst layer attached to each side of the polymeric electrolyte membrane. According to the charge carrier of the polymeric electrolyte membrane, DAFCs fall into two categories: proton electrolyte membrane direct alcohol fuel cells (PEM-DAFCs) using proton electrolyte membrane (PEM) as electrolyte membrane, and anion electrolyte membrane direct alcohol fuel cells (AEM-DAFCs) using anion electrolyte membrane (AEM) as electrolyte membrane. The PEM-DAFCs work in an acidic environment using aqueous solution of alcohol as fuel, while the AEM-DAFCs work in a basic environment using the mixture of alcohol and sodium hydroxide as fuel.

The advantages and disadvantages of PEM-DAFCs and AEM-DAFCs are all related to the electrolyte membranes: PEMs possess much higher chemical stability and ionic conductivity than AEMs, but the electrode kinetics under acidic conditions is much slower than that under basic conditions. Though PEMs and AEMs could share the same backbones in their molecular structures, their functional groups and conducting mechanisms are different, and AEMs are more easily to be attacked by hydroxide anion at high pH and thus are more unstable than PEMs. In addition, both PEM-DAFCs and AEM-DAFCs are impeded by alcohol crossover, the permeation of alcohol across the electrolyte membrane. Another drawback of AEM-DAFCs is the accumulation of pH gradient in the MEA, resulting in an electromotive force in opposite to the fuel cell electromotive force. In this chapter, we limit our discussion to the proton, alcohol, and water transport mechanisms and the rational design of PEMs.

5.2 Types of PEMs

Any PEMs that are applied to DAFCs must possess all the characteristics that are applied to PEMFCs, including high proton conductivity, excellent chemical and electrochemical stability, and adequate mechanical strength and dimensional stability, as well as low alcohol crossover. According to their composition, PEMs could be divided into fully and partially fluorinated polymers and non-fluorinated polymers. The fully and partially fluorinated polymers, such as poly(perfluorosulfonic acid) (PFSA) and sulfonated polymers or copolymers based on α,α,β-trifluorostyrene, possess excellent chemical and electrochemical stability in oxidative and reductive environments and high proton conductivity. However, their conductivities degrade at low humidity or high temperature condition. In addition, they are also limited by the relatively low mechanical stability, high alcohol crossover, high price, and environmental considerations regarding their productions and disposals. The non-fluorinated polymers are mainly based on sulfonated aromatic polymers, such as poly(ether ether ketone) (SPEEK), sulfonated poly(ether sulfone) (SPES), sulfonated poly(styrene) (SPS), and sulfonated polyimide (SPI). The advantages of non-fluorinated polymers include their low price, good mechanical stability, and easiness of synthesis. The disadvantages of sulfonated non-fluorinated polymers include their relatively low chemical and electrochemical stability and proton conductivity degradation at low humidity or high temperature because the ionization of sulfonic acid group depends on the existence of water. Recently, non-sulfonated materials such as phosphoric acid-doped poly(benzimidazole) (PBI) and phosphonic acid based polymers are attracting more and more research attentions attributing to their independence on the existence of water.

Among all the proton conducting materials, the PFSA possesses the best overall performances except high-temperature conductivity and is the most widely accepted proton conducting material for PEMFCs operated at temperatures below the boiling point of water. However, the PFSA possesses inadequate alcohol crossover, as high as 9×10^{-7} $cm^{-2}{\cdot}s^{-1}$, and thus unmodified PFSA is inadequate for DAFCs. In order to improve the overall performances of PFSA, the PFSA is usually modified by doping of a secondary material or by structural modification. For PEMFC application, the intention of modification of PFSA is to improve its water retention and thus proton conductivity at high-temperature and to increase the fuel cell operational temperature. For DAFC application, the fuel cell operational temperature is usually in the liquidus temperature of alcohol below the boiling point of water and thus its high-temperature performances are irrelevant. The bottleneck of PFSA for DAFC application is alcohol crossover, and the target of modification of PFSA is to decrease the alcohol permeation.

The other characteristics of DAFCs that differ from PEMFCs are their service life. For any PEMFCs used as power sources for vehicles, the fuel cells are expected to operate for more than ten years or an equivalent of 10,000 hours, while for DAFCs for portable consumer electronics, short service life is acceptable because of the fast upgrade of consumer electronics. Therefore, the most important characteristics that should be concerned in the rational design of PEMs for DAFCs are the transport characteristics of proton, alcohol, and water. For high-temperature fuel cells, proton conductivity at high temperature is the most essential properties that should be concerned.

5.3 Theoretical and Experimental Method for Design of Proton Conducting Materials

5.3.1 Design of Proton Conducting Materials

Materials design is defined as the ability to predict and tailor the properties and responses of a material for a given engineering application by controlling its chemistry and processing history. Thanks to the great development and large scale installation of modern computing facilities and the development of computational software of ab initio calculations and molecular modeling, materials design has developed into an important research field in the materials science and engineering and attracts more and more interests from both academic institutions and industrial enterprises.

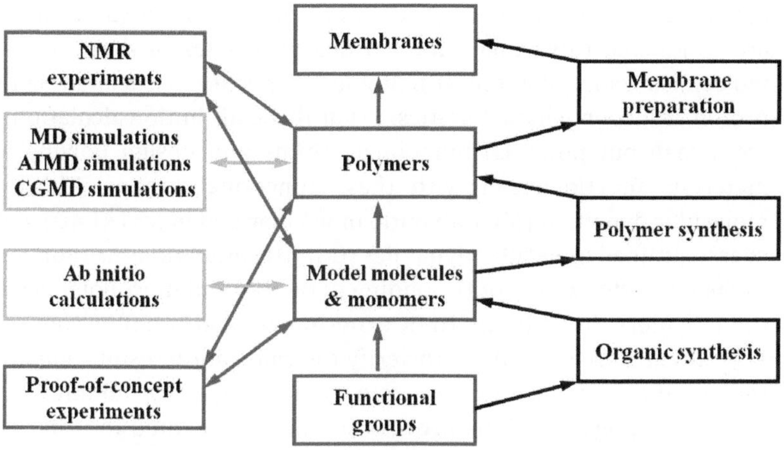

Figure 5.1 Schematic diagram of the rational design of PEMs.

The application of materials design methodologies has accelerated the development of novel materials and reduced the research and development cost. And now, the materials design methodologies are becoming uncompensated measurement and are employed for more or less during the development of any commercial materials [1].

As materials design is the application of all techniques leading to the discovery of new chemical entities with specific properties required for the intended application, we are going to discuss the applications of ab initio calculations and molecular dynamics simulations in the rational design of proton conducting materials. In addition, we will also discuss the applications of NMR techniques and proof-of-concept experiments, which are essential experimental methods for the validation of theoretical calculations.

The schematic diagram of rational design of proton conducting materials is shown in Fig. 5.1. The fundamental building blocks of any proton conducting material is the functional groups which provide the essential properties of the material. Therefore, the first step of materials design is to design or to screen out the functional groups that should be functionalized to the backbone of a polymeric material. On the basis of these functional groups and the knowledge of organic synthesis, we design model molecules which are usually small organic molecules and are available as commercial products or reagents. And then, ab initio calculations are conducted to calculate the various properties of these model molecules, especially the H-bonding properties associated with proton transport characteristics. From these ab initio calculations, we screen out potential functional groups and design polymeric materials functionalized with these functional groups. Thirdly, molecular dynamics (MD), ab initio molecular dynamics (AIMD) and coarse-grained molecular dynamics (CGMD) simulations are applied to the characterization of H-bonding networks and morphologies of the polymers. In addition, NMR experiments and proof-of-concept experiments are carried out to verify the calculation results. Finally, the polymers are synthesized from monomers or by modification of existing polymers. This process should be repeated in order to obtain the best performances of the proton conducting materials.

In the following subsections, we are going to discuss the applications of molecular design methodologies in the development of proton conducting materials and the study of transport mechanism in PEMs.

5.3.2 Transport Mechanisms

The performances of a DAFC depend largely on the transport of proton, alcohol, and water in the PEM. And their transport mechanisms depend not only on the proton conducting material but also on the operational condition of the DAFC. This section will discuss the transport mechanisms of proton, alcohol, and water in PEMs.

The protons transport via vehicle and/or Grotthuss mechanism (molecular and/or structural diffusion mechanism). In the vehicle mechanism, the protons are associated with water molecules forming hydronium ions H_3O^+, Zundel cations $H_5O_2^+$, and Eigen cations $H_9O_4^+$ and these cations transport as a whole entity. While in the Grotthuss mechanism, the protons hop from one site to the other site involving a periodic series of isomerizations between Eigen cations and Zundel cations without the transport of any water molecules. In aqueous solution of acid, the protons transport via mixed vehicle and Grotthuss mechanisms [2, 3]. In hydrated PEMs, the protons also transport via mixed vehicle and Grotthuss mechanisms, while in anhydrous PEMs, the Grotthuss mechanism is the only possible transport mechanism. In anhydrous PEMs, a pervasive H-bonding network should be formed and the protons transport from H-bonding donor to acceptor without interruption. If the H-bonding network is not pervasive, the proton transport pathways are interrupted and tortuous and the proton transport will be inefficient.

The hopping mechanism is only applicable to the transport of protons and not to the transport of the water molecules and alcohol molecules. These neutral molecules can transport only via diffusion mechanism. The other transport mechanism of neutral molecules is the electroosmotic drag, where the neutral molecule is associated with a transporting proton. When the electric field drags the proton, the neutral molecule accelerates simultaneously because of the association interaction.

In the following subsections, we will discuss the three major theoretical and experimental methods that are widely applied to the study of H-bonding network and proton transport and transport of water molecules and alcohol molecules including the ab initio calculations, especially the density functional theory (DFT) calculations, MD, AIMD and CGMD simulations, and NMR techniques.

5.3.3 Ab initio Calculations

Ab initio calculations, especially DFT calculations, have been widely accepted for the study of molecular structures, proton transport mechanism, and H-bonding structure [4].

We studied the acid–base equilibrium [5], the rotational potential and barrier of phosphonic acid groups [6], the H-bonding in acid–base complexes and acid–base amphoteric molecules by DFT calculations or combined with ^{1}H NMR and ^{31}P NMR spectroscopy [7, 8]. Paddison et al. compared the H-bonding characteristics of sulfonic acid, phosphonic acid, and imidazole functionalized alkanes [9], and studied proton transport in phosphonic acid functionalized molecules [10], and in phosphoric acid clusters by DFT calculations [11]. Pereira studied intermolecular interactions and formation of the hydration sphere in phosphonic acid model systems to describe poly(vinyl phosphonic acid) (PVPA) systems by DFT calculations

[12]. Glezakou applied the ab initio calculations to study the acid/base equilibria in clusters [13]. Blanchard et al. studied H-bonding characteristics in phosphonic acid by DFT calculations combined with NMR spectroscopy [14]. Zhou et al. studied the intra- and intermolecular proton transfer in 1*H*-1,2,3-triazole based systems by DFT calculations [15], Nagamani et al. compared the proton transport of 1*H*-1,2,3-triazole, imidazole, and pyrazole [16, 17], and evaluated the importance of dynamic H-bonding and reorientation barriers in proton transport using DFT calculations [18]. Mangiatordi et al. modeled the proton transport in imidazole-like dimers [19], evaluated the Grotthuss proton transport mechanism and proposed a new proton transport mechanism in poly(4-vinyl-imidazole) [20], and the backbone effect of poly(4-vinyl-imidazole) by DFT calculations combined with MD simulations [21]. In addition, Wu et al. studied H-bonding and structural isomerism in protonated water-methanol clusters [22, 23]. Mane studied zero-dimensional nitrogen-based cagelike materials as dopant in PEMs using DFT and the second-order Möller Plesset (MP2) perturbation theory calculations [24].

5.3.4 Molecular Dynamics Simulations

The MD simulations are powerful method to study the formation and rearrangement of H-bonding network in acids and PEMs. For example, Elliott et al. studied the morphology and proton transport in PFSA by MD simulations combined with DFT calculations [25], Hristov studied proton transfer in short-side-chain PFSA by MD simulations [26], Harvey studied H-bonding clustering and phase behavior of imidazole oligomers by MD simulations [27], and Zhu and Yan et al. studied the microscopic structure and H-bonding network of the pristine and phosphoric acid doped PBI [28]. In addition, Li et al. studied proton transfer in water and model polymer electrolyte systems [29], Tsuchida studied proton transport in aqueous sulfuric acid [30] and model systems using MD simulations [31].

The classical MD methods cannot simulate the break and formation of chemical bonds, and thus cannot simulate the Grotthuss proton transport. In this aspect, the AIMD simulations are powerful tools for the direct simulation of bond break and formation. For example, Tsuchida et al. simulated directly the H-bonding network and proton transport in phosphoric acid using AIMD simulations [30], Hayes studied proton transport in triflic acid pentahydrate via ab initio path integral MD simulations [32], Ludueña et al. studied the proton transport in PVPA by use of a steered AIMD approach [33]. The other MD simulation method used for the PEMs is the CGMD simulations which are much faster than the MD and AIMD [34].

In order to evaluate the electroosmotic drag of water and methanol molecules in electric field, we developed a MD method to calculate macroscopic transport velocity of molecules from molecular velocities recorded during MD simulations and compared the average molecular velocities with peak shifted molecular velocities based on molecular velocity distribution functions in electric field [35–37]. We also studied the intermolecular momentum transfer in poly(perfluorosulfonic acid) membrane hydrated by aqueous solution of methanol via velocity auto-correlation functions and velocity cross-correlation functions [38].

For more details about application of ab initio calculations and MD simulations in PEMs, the readers are referred to recent reviews [25, 39, 40].

5.3.5 NMR Techniques

The conventional nuclear magnetic resonance (NMR) techniques show how the resonance frequency of a magnetic nucleus is affected by its electronic environment and the presence of magnetic nuclei in its vicinity, and thus are widely applied to the elucidation of structures of organic molecules. The modern NMR techniques have turned to use pulses of electromagnetic radiation to excite magnetic nuclei or spins and to monitor the radiation of the magnetic spins as they return to equilibrium. These NMR techniques have not only allowed the structural analysis of complex molecules and molecular systems in modern organic chemistry and biochemistry, but also the elucidation of dynamics in complicated molecular systems.

Generally, homogeneous and diamagnetic liquid samples yield sharp signals in NMR spectroscopy attributing to the rapid isotropic molecular tumbling. However, solid samples would broaden the spectrum to the extent that it would be unobservable with a high-resolution NMR spectrometer because of the chemical shift anisotropy and homonuclear dipolar coupling. Such broadening effects can be eliminated by a combination of high-power multiple pulse decoupling and fast spinning around the magic angle at 54.74°. The development of magic angle spinning nuclear magnetic resonance (MAS NMR) has facilitated its application in solid state materials especially in the field of polymeric materials [41, 42]. Today, various NMR techniques, including electrophoretic NMR, pulse gradient spin echo and stimulated echo, high-pressure magic angle spinning, and inversion recovery NMR, and magnetic resonance imaging, have been applied to the characterization of structural and dynamic properties of PEMs [43].

5.3.5.1 H-bonding and proton transport

The proton transport processes depend either on the ionization of acidic moieties or on the formation of pervasive H-bonding networks between H-bonding donors and acceptors. The NMR techniques could provide unique information to the study of H-bonding networks and proton transport processes, including structural and thermodynamics of the H-bonding networks as well as activation energies for the proton transport between H-bonding donors and acceptors, and to the design of proton conducting materials. The important magnetic nuclei being probed in NMR experiments include the ^{1}H nuclei which are directly related to the H-bonding, and the ^{15}N, ^{13}C, or ^{31}P nuclei which are indirectly related to the H-bonding [44]. And the liquid NMR and MAS NMR techniques could be applied to the model systems and the solid polymeric materials, respectively.

Recently, we applied the ^{1}H and the ^{31}P NMR spectra to characterize H-bonding characteristics of acid–base complexes of butanephosphonic acid (**C4PA**) and 1,2,3-triazole, pyridine, pyridazine, pyrimidine, pyrazine, 1,3,5-triazine, indole, 1*H*-benzimidazole, and 1*H*-indazole, and found that there exist fast proton exchange between **C4PA** and 1*H*-benzimidazole and 1*H*-indazole. In addition, the variable temperature ^{1}H NMR experiments further proved that the fast proton exchange is restrained at low temperature about −100°C for acid–base complex of **C4PA** and 1*H*-indazole [7, 8]. We also studied the acid–base equilibrium in various acid and 1*H*-benzimidazole complexes by ^{1}H NMR and DFT calculations [5]. Schechter carried out liquid ^{1}H NMR and ^{31}P NMR measurements of phosphoric acid doped PBI under controlled humidity and showed the existence of free or mobile acid and evaluated the distribution of phosphorus species H_3PO_4, $H_2PO_4^-$ and $H_4P_2O_7$ [45]. Wasmus studied the H_3PO_4-equilibriated NAFION and concluded the existence of fast proton exchanges between H_3PO_4, $-SO_3H$, and residual water [46]. Toppet observed directly the fast prototropic rearrangement in poly(4-vinyltiazole) by ^{13}C NMR [47].

MAS ^{1}H NMR is a powerful tool for the study of H-bonding in solid polymeric materials because the proton isotropic chemical shift exhibits a dependence on the interatomic distances associated with the H-bonding. Goward et al. applied the variable temperature MAS ^{1}H NMR experiments as well as two dimensional (2D) homonuclear double quantum NMR and 2D exchange spectroscopy to the ethylene oxide tethered imidazole heterocycles [48]. Double quantum NMR experiments could identify those hydrogen-bonded protons that are mobile on the time scale of the experiment and, thereby, those that are able to participate in charge transport. Blanchard et al. applied the MAS ^{1}H NMR combined with DFT calculations to the study of hydrogen-bonded phosphonic acids and resolved multiple H-bonding sites in a set of crystalline phosphonic acids that contain various functional groups, including $-COOH$, $-PO_3H_2$, and $-NH_3^+$ [14]. Çelik and Bozkurt et al. detected two different types of hydrogen-bonded protons in the poly(VPA-*co*-VTri) at anhydrous state by MAS ^{1}H NMR [49], and structural features and molecular dynamics in blend of PVPA and ABPBI by ^{1}H, ^{13}C, and ^{31}P MAS NMR as well as variable temperature ^{1}H MAS and one-dimensional double-quantum NMR [50]. The MAS NMR techniques are also applied to the study of acid–base poly(acrylic acid)-poly(4-vinyl pyridine) polymer blend [51], triazole-functionalized siloxane polymers [52], poly(*p*-vinylbenzylphosphonic acid) [53], PVPA [54, 55], imidazole-based proton conducting materials [48, 56–59], polyvinyl alcohol [60], poly(*p*[2,2′-(*m*-phenylene)-5,5′-bibenzimidazole]) [61], NAFION and sulfonated poly(ether ether ketone) [62]. Other magnetic nuclei are also detected in acid–base complexes of heterocyclic and dicarboxylic acids using ^{15}N NMR [63] and ^{2}H NMR [64].

5.3.5.2 Diffusion coefficients

The experimental measurement of self-diffusion coefficient D requires the introduction of radioactive or stable isotopes or fluorescence markers to label the observed molecules and the detection of time and space dependence of concentration of the labeled species, and the fitting of experimentally observed time and space dependence of concentration to the solution of diffusion equation with the associated initial and boundary conditions with D as adjustable parameter. Modern NMR techniques based on the pulse gradient spin echo and pulse gradient stimulated echo pulse sequences can provide very accurate and quick measurement of D in multicomponent systems without the need of isotopic labeling attributing to the chemical shift difference of the diffusing species [43].

Zawodzinski et al. applied the pulsed field gradient spin-echo ^{1}H NMR to the determination of water diffusion coefficients in perfluorosulfonate ionomeric membranes [65]. Every et al. applied a modified pulsed field gradient ^{1}H NMR method to determine the methanol diffusion in NAFION 117 and in a 40% disulfonated wholly aromatic poly(arylene ether sulfone). This method allows the determination of methanol diffusion coefficient within the membrane while immersed in a methanol solution by using a second set of gradient pulses to suppress the signal from the methanol in solution and thus allowing the methanol within the membrane to be monitored unambiguously [66]. The pulsed field gradient NMR techniques have also been used to evaluate diffusion coefficients of water and methanol in NAFION [67, 68], NAFION composite [69, 70], sulfonated poly(arylene sulfone) membranes [66, 71], cation exchange copolymers [72] cross-linked poly(vinyl alcohol)-poly(vinyl pyrrolidone) membranes [73], and sulfonic acid-grafted cellulose membranes [74], sulfonated poly(4-phenoxybenzoly-1,4-phenylene) [67], ion transport in phosphoric acid doped PBI membranes [75], proton diffusion in phosphoric acid doped PBI

[76], and anion electrolyte membranes [77, 78] and heterogeneous solvent diffusion environments in anion exchange membranes [79]. The pulsed field gradient NMR techniques are also applied to evaluate diffusion coefficients of chemically different protons in sulfonic acid, phosphonic acid and imidazole functionalized model compounds [80]. For detailed information about the NMR techniques for the study of ion and molecular transport in polymer electrolytes, the readers are referred to a recent review [81].

In addition, high-pressure NMR diffusion measurements were also carried out in a static field gradient to get supplemental information regarding water transport in SPTES [71], electrophoretic NMR was applied to the measurement of electroosmotic drag coefficients K_d in polymer electrolyte membranes [82], and magnetic resonance micro-imaging techniques to the microscopic states of water and methanol in NAFION [83].

5.3.6 Proof-of-Concept Experiments

The final step before the experimental synthesis of the proton conducting materials and the preparation of PEMs is the proof-of-concept experiment. During the proof-of-concept experiment, model molecules with adequate functional groups are bought or synthesized, and model molecular systems are prepared from these model molecules and their proton conductivities are measured under various conditions. Recently, we carried out proof-of-concept experiment by measurement of proton conductivity of acid–base complex of methyl phosphonic acid and $1H$-1,2,3-triazole, and found that the conductivity increases for about 50 folds when 10% of methyl phosphonic acid was added to $1H$-1,2,3-triazole, and verified the results concluded from DFT calculations and ^{1}H NMR spectroscopy experiments [8].

5.4 An Example

Acid–base polymeric complexes and acid–base amphoteric polymers are potential high-temperature proton conducting materials considered to be potential candidate to the high-temperature fuel cells operated at temperatures well above the boiling point of water since these materials exhibit promising proton conducting properties under dehydrated or even anhydrous states. These materials are usually composed of acidic phosphonic acid groups and nitrogenous heterocyclic groups, and their proton conducting properties depend on the formation and rearrangement of pervasive H-bonding network between the acidic and the basic groups. In this section, we will introduce an example about the design of acid–base proton conducting materials, especially the screening of basic functional groups from a couple of candidate nitrogenous heterocycles using the procedure proposed in the previous section.

5.4.1 Model Molecules

In order to screen out suitable basic functional groups for the high-temperature proton conducting material based on acid–base polymeric complex or amphoteric polymer, a couple of nitrogenous heterocyclic bases are designed as model molecules and their H-bonding interaction with the acidic model molecule **C4PA** will be studied using ab initio calculations and ^{1}H and ^{31}P NMR experiments, and the proton conductivity of model molecular systems will be verified by proof-of-concept experiments.

The nitrogenous heterocyclic bases are composed of one 6-membered ring with one to three nitrogen atoms, or fused benzene ring and 5-membered ring with one or two nitrogen atoms as shown in Fig. 5.2. The corresponding acid–base complexes of **C4PA** and nitrogenous heterocyclic bases **B-1** to **B-11** will be nominated as **C-1** to **C-11**, respectively, in this section. These nitrogenous heterocyclic moieties are representatives of various basic ingredients which could be found in high-temperature PEMs as heterocyclic solvents and basic polymers and copolymers. Therefore, the H-bonding and proton transport characteristics revealed from these model systems are representative to the relevant acid–base complexes.

5.4.2 DFT Calculation Method

All the calculations were carried out based on DFT and single-point MP2 perturbation theory as implemented in the GAUSSIAN

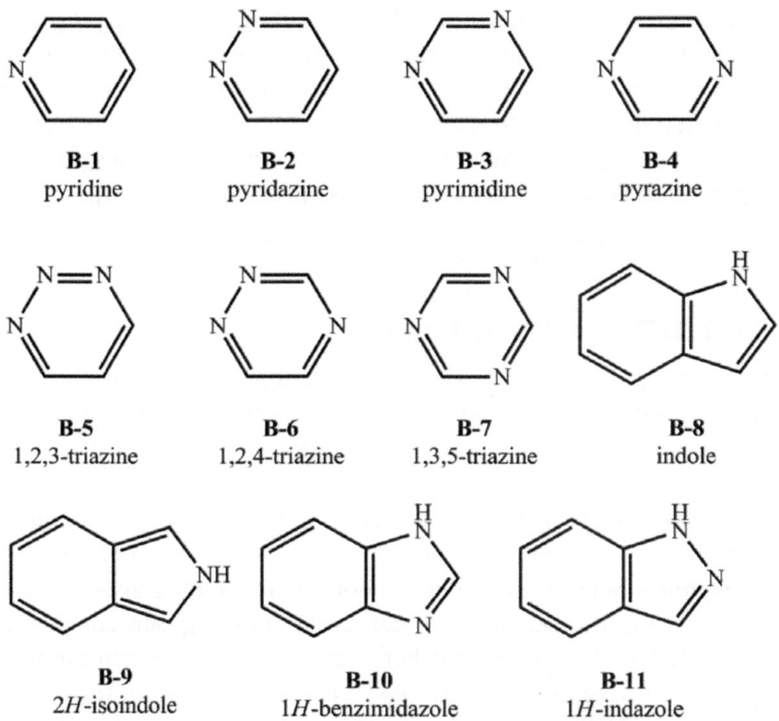

Figure 5.2 The model molecules of nitrogenous heterocycles.

03 package [84]. Firstly, all the acid and base molecules and the acid–base complexes were optimized at the B3LYP/6-311+G(2d,p) level of theory. And then, vibrational frequency calculations were conducted to verify if local minima had reached (with or without imaginary frequencies). If local minimum of a molecular system did not reach, the molecular system was adjusted and reoptimized until the reach of local minimum. Finally, single-point calculations were performed and energies were corrected at the MP2/6-311+G(2d,p) level of theory for all the optimized molecular systems.

The intermolecular H-bonding energy of an acid–base complex was calculated as the energy difference between the acid–base complex and the acid and the base molecules [85, 86],

$$\Delta E_{\mathrm{HB}} = E_{\mathrm{AB}} - E_{\mathrm{A}} - E_{\mathrm{B}} + \mathrm{ZPE} \tag{5.1}$$

where ΔE_{HB} is the intermolecular H-bonding energy, E_{AB}, E_{A}, and E_{B} are total energies of the acid–base complex and the acid and the base molecules, respectively. ZPE is the zero-point energy correction evaluated during the frequency calculation. The intermolecular H-bonding energy is further corrected by the basis set superposition error (BSSE) based on the counterpoise procedure [87, 88],

$$\Delta E_{\mathrm{HB,\,C}} = E_{\mathrm{AB}} - E_{\mathrm{A}} - E_{\mathrm{B}} + ZPE + BSSE \qquad (5.2)$$

5.4.3 DFT Calculation Results*

The H-bonding structures and parameters of the acid–base complexes calculated at the B3LYP/6-311+G(2d,p)//MP2/6-311+G(2d,p) level of theory were summarized in Table 5.1. From the structures of optimized acid–base complexes, it is concluded that there are two types of H-bonding, the $\mathrm{N} \cdots \mathrm{H} - \mathrm{O}$ bonding and $\mathrm{O} \cdots \mathrm{H} - \mathrm{N}$ bonding, formed in the acid–base complexes. The phosphonic acid group could act both as H-bonding acceptor and donor, while the secondary heterocyclic amine group could only act as H-bonding donor and heterocyclic imine nitrogen atom as H-bonding acceptor. The strongest H-bonding with a H-bonding energy at -17.7 kcal·mol^{-1} is formed in the complex of **C4PA** and 1*H*-indazole (**B-11**) as both **C4PA** and 1*H*-indazole (**B-11**) could act simultaneously as H-bonding donor and acceptor and double H-bonding (one $\mathrm{N} \cdots \mathrm{H} - \mathrm{O}$ bonding and one $\mathrm{O} \cdots \mathrm{H} - \mathrm{N}$ bonding) are formed. However, all the other complexes could only form single H-bonding thus weaker than complex of **C4PA** and 1*H*-indazole (**B-11**). When the **C4PA** forms H-bonding with a secondary amine group in heterocyclic ring, the amine group donates its proton to **C4PA** instead of accepting proton from **C4PA** (the $\mathrm{O} \cdots \mathrm{H} - \mathrm{N}$ bonding) as the amine nitrogen atom has a planar sp^2 hybrid and its lone pair is part of the aromatic system. On the other hand, if **C4PA** forms H-bonding with planar sp^2 nitrogen atom without a hydrogen atom bonded (the imine nitrogen atom), the **C4PA** acts as H-bonding donor and the imine nitrogen atom as H-bonding acceptor forming a $\mathrm{N} \cdots \mathrm{H{-}O}$ bonding. The H-bonding distance

*Adapted with permission from Xie el al. [7] Copyright (2013) American Chemical Society.

Table 5.1 Intermolecular H-bonding parameters for the acid–base complexes

Systems	$d_{N(C)\text{-}H}$ Å	$d_{O\text{-}H}$ Å	∠N(C)HO deg.	υ cm^{-1}	ΔE_{HB} $kcal{\cdot}mol^{-1}$	$\Delta E_{HB,c}$ $kcal\ mol^{-1}$
C-1	1.685	1.014	176.7	2865	−13.0	−12.3
	1.085	2.355	141.8	3162		
C-2	1.722	1.008	173.2	2980	−13.2	−12.6
	1.086	2.261	143.2	3161		
C-3	1.739	1.004	177.0	3056	−10.7	−10.1
	1.086	2.384	140.0	3176		
C-4	1.745	1.003	175.4	3067	−11.6	−11.0
	1.086	2.318	141.4	3160		
C-5	1.784	1.000	171.3	3169	−12.1	−11.4
	1.087	2.199	143.4	3152		
C-6	1.783	0.998	171.9	3167	−12.1	−11.5
	1.086	2.222	142.5	3154		
C-7	1.973	0.996	175.1	3204	−9.7	−9.1
	1.086	2.295	140.2	3171		
C-8	1.016	1.969	151.8	3497	−10.6	−9.8
C-9	1.019	1.965	148.3	3462	−10.2	−9.5
C-10	1.691	1.012	173.0	2905	−14.7	−14.0
	1.082	2.270	134.2	3218		
C-11	1.025	1.836	152.2	3343	−18.5	−17.7
	1.715	1.008	171.1	2972		

of O···H in complexes of **C4PA** and indole (**B-8**) and of **C4PA** and 2*H*-isoindole (**B-9**) are at 1.991 Å and 1.965 Å, much longer than the H-bonding distance of N···H at about 1.74 Å. The H-bonding angles also show the same tendency as the O···H–N bonding usually has an angle at about 150°, while the N···H–O bonding has an angle at about 175°. Besides the H-bonding, the structure of acid–base complexes are also influenced by the weak O···H–C interaction, where the carbon atom is directly bonded to an imine nitrogen atom, with an O···H distance at about 2.199 to 2.384 Å. Therefore, it could be concluded that the structure of acid–base complexes are determined by three types of interactions, the N···H–O bonding, O···H–N bonding, and O···H–C interaction, and the N···H–O bonding is stronger than the O···H–N bonding, and the O···H–C interaction is the weakest.

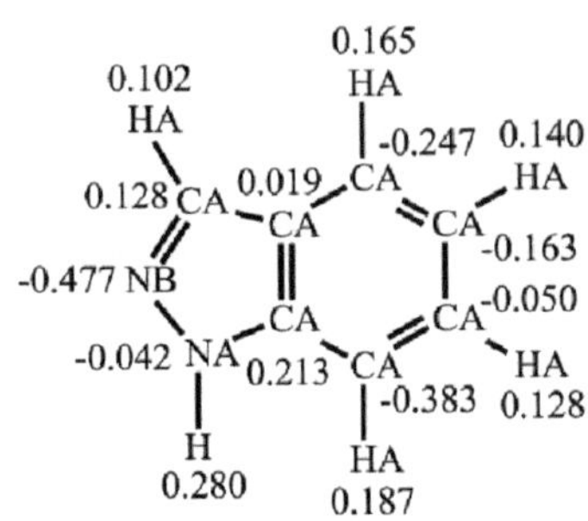

indazole (**B-11**)

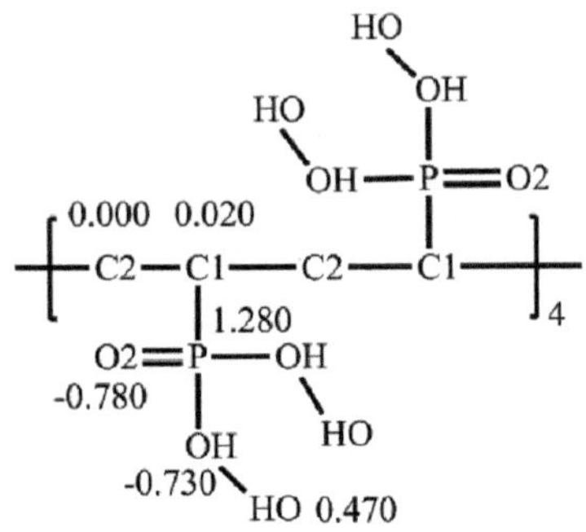

PVPA

Figure 5.3 Site names and residual charges of the model molecules.

These results are consistent with the acid–base complexes of **C4PA** and heterocycles of 5-membered ring, where the complex of **C4PA** and pyrazole forms one O · · · H−N bonding and one N · · · H−O bonding, the complex of **C4PA** and imidazole forms a strong N · · · H−O bonding, and the complex of **C4PA** and pyrrole forms a weak O · · · H − N bonding [89].

5.4.4 MD Simulations

In this section, we will give MD simulation examples for indazole (**B-11**), PVPA, and their complexes. In our MD simulations, mixed force field models were applied: The indazole (**B-11**) was modeled using allatom force field model and PVPA was modeled using united-atom force field model with hydrogen atoms which are bonded to carbon atoms in the main chain modeled implicitly. In Fig 5.3, it shows the site names of indazole (**B-11**) and PVPA, as well as the residual charge of each atom or united atom evaluated from DFT calculations at theory level of B3LYP/6-311+G(d,p) using the ChelpG algorithm as implemented in the GAUSSIAN 03 package [84]. The DREIDING generic force field model was used for both molecules [90], and the optimized bond lengths, bond angles and dihedral angles from DFT calculations were applied as reference bonding parameters in the force field model.

The MD simulations were carried out using the DL_POLY program [91]. 3-D periodic boundary conditions were applied to the simulation cells, and the simulations were carried out in the NPT ensemble with temperature and pressure maintained, respectively, by the Nosé-Hoover thermostat and barostat with the same relaxation time parameters of 0.2 ps [92, 93]. The Verlet integration scheme was used to solve the Newtonian equations of motion for atoms or united atoms in the simulation cells with an integration step time of 1 fs [94]. The initial simulation cells were constructed by insertion of the molecules randomly. During the initial simulations, the loose initial simulation cells were gradually squeezed and reached equilibrium densities. After reaching of the equilibrium densities, another 1,000,000 steps were carried out for the generation of structural characteristics of the systems.

In Fig. 5.4, it shows snapshots of the model systems recorded during MD simulations. From these snapshots, it could be seen that the molecules distribute in the simulation cells randomly and evenly without any significant order.

In Fig. 5.5, it shows the radial distribution functions $g(r)$ and integrated radial distribution functions $n(r)$, which are related to the H-bonding network, evaluated from the MD simulations. For the indazole (**B-11**), there is only one type of H-bonding interaction between imine nitrogen atom NB and H, noticing that there is no H-bonding interaction between amine nitrogen atom NA and H. The first peak position of $g_{\mathrm{NB-H}}(r)$ is 2.25 Å indicating significant H-bonding interaction between NB and H, and that for $g_{\mathrm{NA-H}}(r)$ is 3.38 Å indicating negligible H-bonding interaction between NA and H. From the integrated radial distribution functions $n_{\mathrm{NB-H}}(r)$, we got the first coordination number at about 0.68 (value at first valley of $g_{\mathrm{NB-H}}(r)$), which means that there is about 0.68 H-bonding interactions for each NB atom. For the PVPA, there are two types of H-bonding interactions, O2–HO interaction and OH–HO interaction, formed and the O2–HO interaction is more powerful than the OH–HO interaction. The first peak for $g_{\mathrm{O2-HO}}(r)$ is at 1.38 Å and number of H-bonding interactions is about 1.20, while for the OH–HO interaction, the first peak is at 1.53 Å and the number of H-bonding interactions is only 0.36. Since the OH atom has already

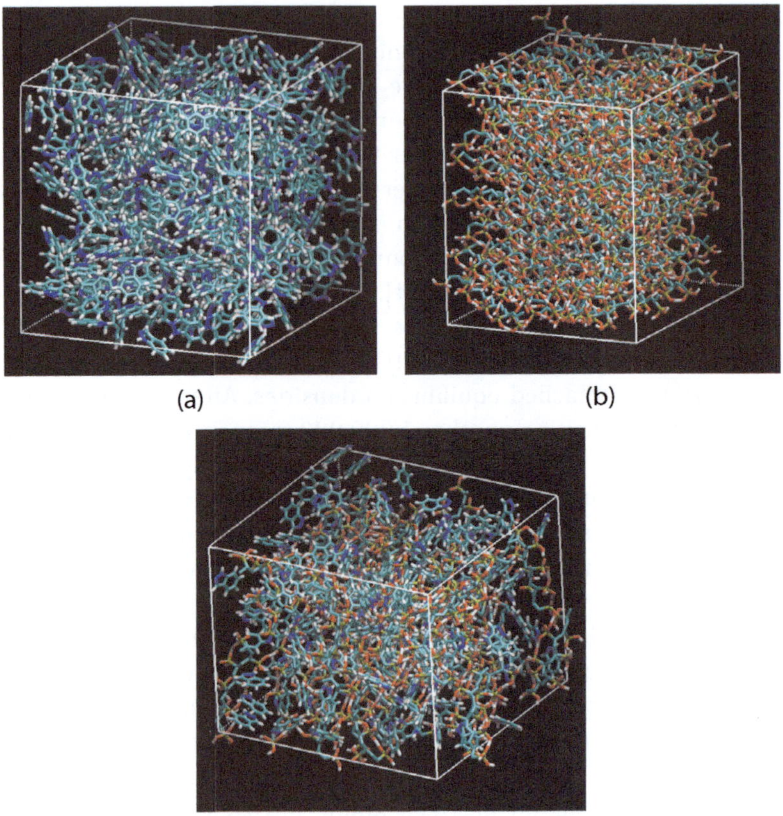

(a) (b) (c)

Figure 5.4 Snapshots of the simulated systems: (a) indazole (**B-11**), (b) PVPA, and (c) complex of indazole (**B-11**) and PVPA (color codes: C cyan, H white, O red, N blue, and P yellow).

one acidic hydrogen atom HO bonded, the overall hydrogen atoms bonded to OH is 1.36, slightly greater than that bonded to O2.

In the complex of indazole (**B-11**) and PVPA, the first peak position for $g_{\mathrm{O2-HO}}(r)$ is at 1.38 Å, the same as that of pristine PVPA, however, the number of H-bonding interactions decreases to 1.06. The other difference between pristine PVPA and complex of indazole (**B-11**) and PVPA is the formation of O2–H interaction, or the H-bonding between indazole (**B-11**) and PVPA. The first peak

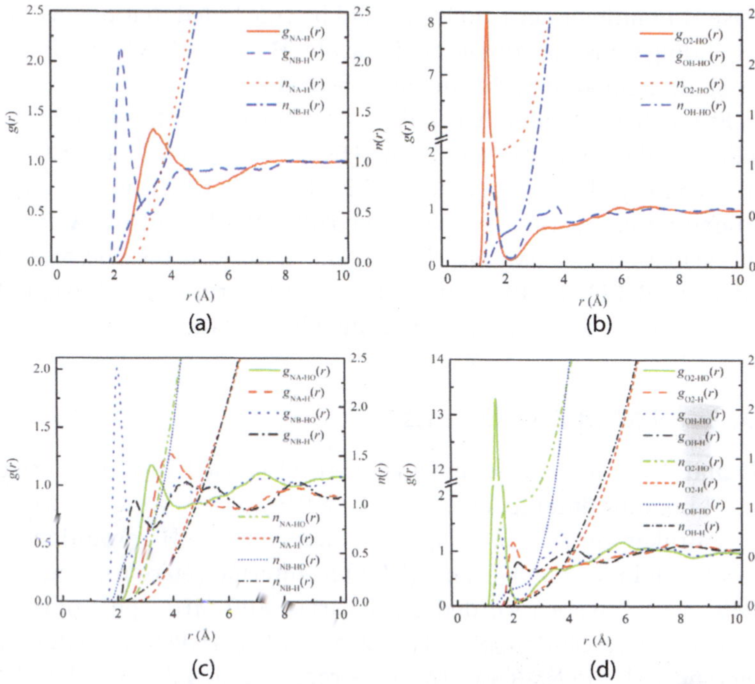

Figure 5.5 Radial distribution functions for (a) indazole (**B-11**), (b) PVPA, and (c, d) complex of indazole (**B-11**) and PVPA.

for $g_{O2-H}(r)$ is at 2.00 Å, and the number of H-bonding interaction is about 0.16. Thus the H-bonding between indazole (**B-11**) and PVPA is much weaker than that between O2 and HO. The overall number of H-bonding of O2 is 2.22, almost the same as that in pristine PVPA. For OH–HO interaction, the first peak position is at 1.53 Å with about 0.20 H-bonding interactions. By adding the number of H-bonding interactions of OH–H at 0.12, the overall H-bonding interactions for OH are almost the same compared to that of pristine PVPA. The H-bonding ability for NA is negligible with first peak position for NA–HO at 3.23 Å and that for NA–H at 3.80 Å. The NB–H interaction distance is at 2.63 Å and the number of interactions is about 0.18, much weaker than that in indazole (**B-11**). The H-bonding distance for NB–HO is about 1.95 Å, and the number of interactions at 0.60, much stronger than that between NB and H. By adding up these H-bonding interactions, it could be concluded that the overall H-bonding for NB in complex of indazole (**B-11**) and PVPA is almost the same compared to that in indazole (**B-11**).

From these results, it could be concluded that the H-bonding ability in the complex of indazole (**B-11**) and PVPA is almost the same compared to that in indazole (**B-11**) and in PVPA, however, the H-bonding interactions between indazole (**B-11**) and PVPA cause the formation of pervasive H-bonding network in the complex of indazole (**B-11**) and PVPA and thus ease the structural diffusion of protons under dehydrated or even anhydrous state.

5.4.5 ^{1}H NMR Experiments*

^{1}H and ^{31}P NMR measurements were also used to characterize the H-bonding interactions. The ^{1}H NMR spectra were recorded on a Bruker AV 500 MHz spectrometer with a ^{1}H resolution of 0.45 Hz and tetramethylsilane (TMS) as internal reference, and ^{31}P NMR spectra were recorded on a Varian 400 MHz spectrometer with a ^{31}P resolution of 2.47 Hz and 85% H_3PO_4 as external reference. All the NMR spectra were recorded in solvent of DMSO-d_6 (99.9%, containing 0.03% v/v TMS, supplied by the Cambridge Isotope Laboratories, Inc.) The **C4PA** (98%) was supplied by Alfa Inc., pyrimidine (**B-3**) (97%) and 1*H*-benzimidazole (**B-10**) (99%) were supplied by the Aladdin Inc., pyridine (**B-1**) was supplied by Sinopharm Inc., 1*H*-indazole (**B-11**) (99%), pyridazine (**B-2**) (98%), pyrazine (**B-4**) (99%), and indole (**B-8**) (99%) were supplied by Adamas Inc., and 1,3,5-triazine (**B-7**) (98%) was supplied by TCI Inc. All the chemicals were used as received without further purification but were degassed by freeze-pump-thaw cycles before use.

^{1}H NMR chemical shift reflects the chemical environment of a proton since a tightly bound proton is shielded and a loosely bound proton is deshielded by electron cloud. The acidic or transportable protons are usually significantly deshielded with chemical shifts at 8 ppm or greater. In DMSO-d_6, the chemical shift of acidic protons

*Adapted with permission from Xie et al. [7] Copyright (2013) American Chemical Society.

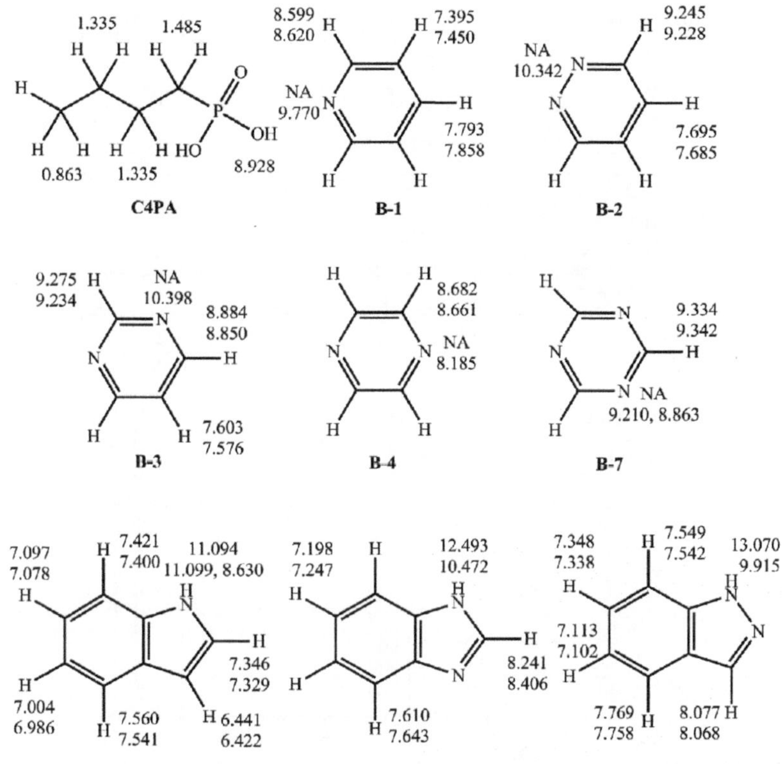

Figure 5.6 ^{1}H NMR chemical shifts (ppm) of the **C4PA** and nitrogenous heterocycles (upper number) and their complexes (lower number).

of **C4PA** is 8.928 ppm (Fig. 5.6). If a heterocyclic base is added and H-bonding is formed between the **C4PA** and the heterocyclic base, the acidic proton will be further deshielded. Therefore, the H-bonding and proton exchange between a **C4PA** and a heterocyclic base could be studied by comparison of ^{1}H NMR chemical shift of **C4PA**, heterocyclic base, and their complex.

Besides the acidic proton of **C4PA**, the heterocyclic bases 1*H*-benzimidazole (**B-10**), 1*H*-indazole (**B-11**), and indole (**B-8**) also possess acidic proton (N-H) with chemical shifts at 12.493, 13.070, and 11.094 ppm, respectively. When 1*H*-benzimidazole (**B-10**) or 1*H*-indazole (**B-11**) is added to **C4PA** forming an acid–base

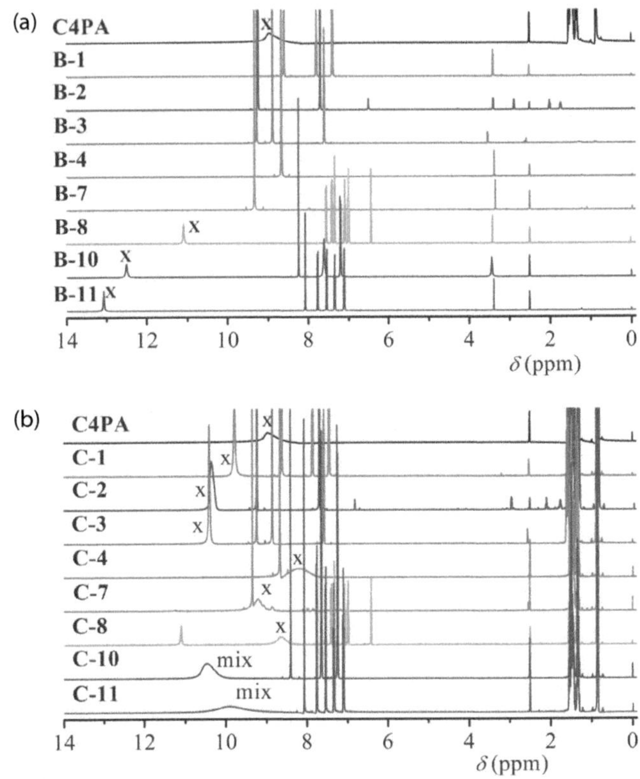

Figure 5.7 ^{1}H NMR spectroscopy of (a) **C4PA** and nitrogenous heterocycles, and (b) their complexes in DMSO-d6. Reprinted with permission from Xie et al. [7]. Copyright © 2013 American Chemical Society.

complex, the two peaks from **C4PA** and heterocyclic base are mixed into a single peak indicating the fast proton exchange between **C4PA** and heterocyclic base (Fig. 5.7). However, if indole (**B-8**) is added to **C4PA** forming an acid–base complex **C-8**, the two peaks from **C4PA** and indole (**B-8**) do not mix into a single peak indicating no proton exchange or proton exchange is slower than time scale of ^{1}H NMR measurements at milliseconds. This characteristic is consistent with the fact that indole (**B-8**), unlike most amines, is an acid with a pK_a of 16.2 (for the N-H proton), and very strong acids such as hydrochloric acid are required to protonate indole (**B-8**) (the protonated form has a pK_a of −3.6) [95]. These characteristics are also consistent with the calculated H-bonding energies of the acid–base complexes, the H-bonding energies for complexes of **C4PA** and 1*H*-benzimidazole (**B-10**), or **C4PA** and 1*H*-indazole (**B-11**), or **C4PA** and indole (**B-8**) are −14.0, −17.7, or −9.8 kcal mol^{-1}, respectively. Therefore, it is concluded that strong H-bonding interaction could cause fast proton exchange.

For the complex of **C4PA** and heterocyclic base without acidic proton, we compared the chemical shift between **C4PA** and its complex with heterocyclic base. For complex of **C4PA** and pyridine (**B-1**), the chemical shift of the acidic proton is shifted downfield 0.842 ppm from 8.928 to 9.770 ppm, and the H-bonding energy of the complex is −12.3 kcal mol^{-1}. For complex of **C4PA** and pyridazine (**B-2**) or **C4PA** and pyrimidine (**B-3**), the chemical shift is also shifted downfield 1.414 or 1.470 ppm, and the corresponding H-bonding energy is −12.6 or −10.1 kcal mol^{-1}, respectively. Different characteristics were observed for complex of **C4PA** and pyrazine (**B-4**) or **C4PA** and 1,3,5-trazine (**B-7**): The ^{1}H NMR chemical shift of complex of **C-4** is shifted upfield 0.753 ppm, and that of **C-7** splits into two peaks one downfield 0.282 ppm and the other upfield 0.065 ppm indicating the existence of protons in different chemical environments thus complicated H-bonding characteristics.

The chemical shift for ^{31}P NMR reflect the chemical environment of the phosphonic acid group, and the H-bond characteristics of the phosphonic acid group in the acid–base complex could be studied by comparison of ^{31}P chemical shifts between an isolated C4PA or its complex with a heterocyclic base. The ^{31}P NMR chemical shift for C4PA is 29.875 ppm, and significant upfield shift was observed in the acid–base complexes (Fig. 5.8). When 1*H*-benzimidazole (**B-10**), a medium strength base, is added to C4PA solution in DMSO-d_6, the ^{31}P chemical shift is shifted upfield 1.873 ppm owing to the attraction of acidic proton of C4PA by 1*H*-benzimidazole (**B-10**), and the leftover electron cloud on the phosphonic acid group that shields the ^{31}P atom. The other heterocyclic bases, such as pyridine (**B-1**), will also attract proton from C4PA and leave electron cloud over the phosphonic acid group as indicated by the upfield shift of the chemical shift for ^{31}P NMR. These characteristics are consistent with the density functional theory calculations in terms of H-bond structures and H-bond energies.

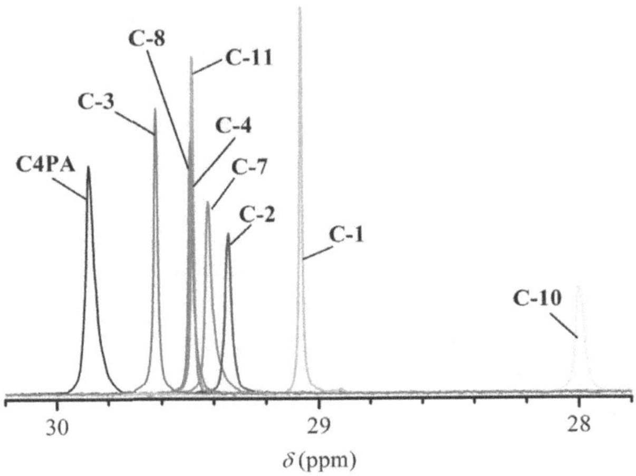

Figure 5.8 ^{31}P NMR spectroscopy of **C4PA** and its complexes with heterocycles in DMSO-d6.

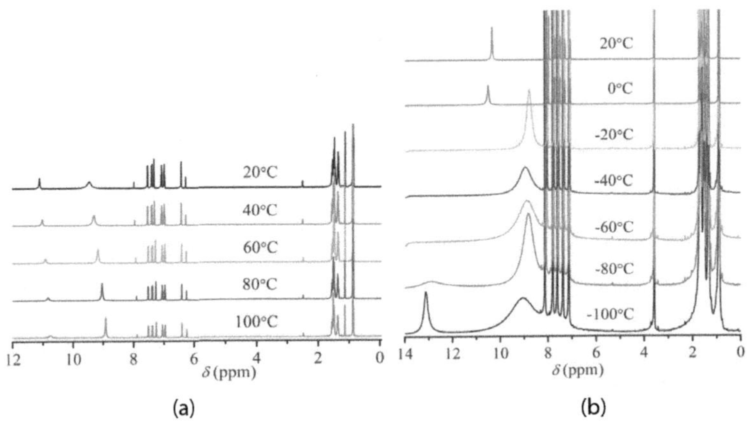

Figure 5.9 Variable temperature ^{1}H NMR spectra of (a) complex **C4PA** and indole (**B-8**), and (b) complex **C4PA** and 1H-indazole (**B-11**). Reprinted with permission from Xie et al. [7]. Copyright © 2013 American Chemical Society.

In Fig. 5.9, it shows the variable temperature ^{1}H NMR spectroscopy of complexes of **C4PA** and indole (**B-8**) and 1H-indazole (**B-11**). The two peaks from **C4PA** and indole (**B-8**) do not merge up to 100°C, indicating no proton exchange, thus no structural proton

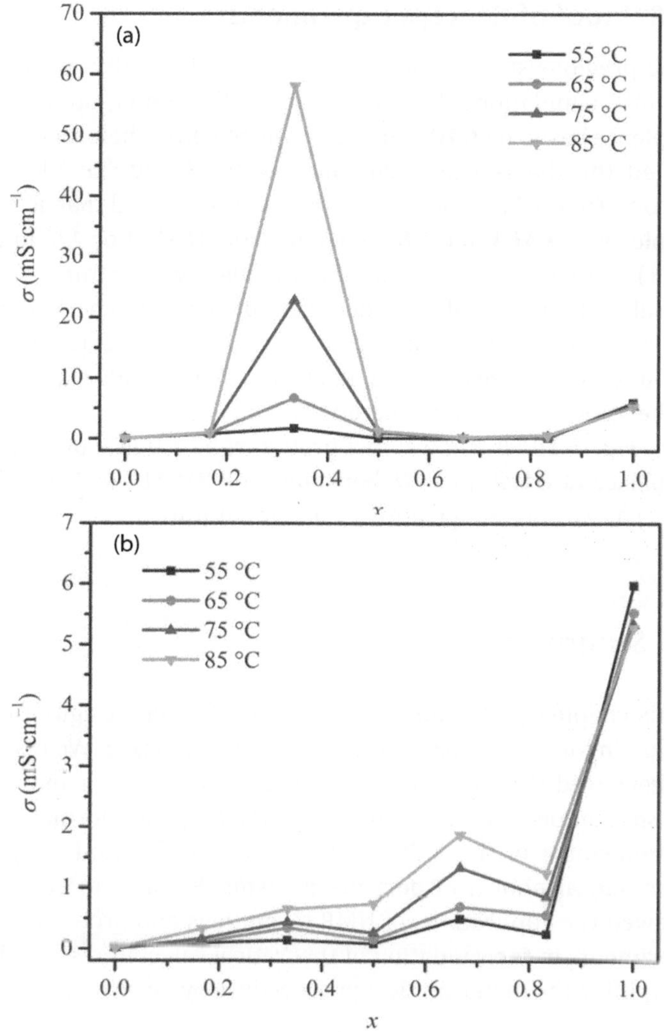

Figure 5.10 Proton conductivities of complexes of **C4PA:** (a) 1*H*-benzimidazole (**B-10**) and (b) 1*H*-indazole (**B-11**).

transport. For the complex of **C4PA** and 1*H*-indazole (**B-11**), the proton exchange between **C4PA** and 1*H*-indazole (**B-11**) is not restrained until −80°C, indicating low potential barrier of structural proton transport.

5.4.6 Proof-of-Concept Experiments

In the previous subsections, we have applied ab initio calculations and MD simulations to the study of H-bonding in acid–base complexes based on **C4PA** and model nitrogenous heterocycles, and verified the theoretical calculations with ^{1}H and ^{31}P NMR spectroscopy. On the basis of this research, we concluded that acid–base complexes of **C4PA** and $1H$-benzimidazole (**B-10**) or $1H$-indazole (**B-11**) are potential candidate for the high-temperature PEMs. In this subsection, we will carry out proof-of-concept experiments, the experimental measurement of proton conductivities of the proposed acid–base complexes, as the final step before the experimental synthesis of the polymeric materials.

In Fig. 5.10, it shows the proton conductivity of acid–base complexes of **C4PA** and $1H$-benzimidazole (**B-10**) or $1H$-indazole (**B-11**) at various compositions and temperatures.

5.5 Summary

In this chapter, we proposed a procedure for the design of proton conducting materials based on acid–base complexes. We discussed and reviewed the applications of ab initio calculations and various MD simulations in the study of H-bonding and formation and rearrangement of H-bonding network and in the study of proton, water, and alcohol transport mechanisms. We also discussed and reviewed the applications of NMR techniques and proof-of-concept experiments in the validation of theoretical calculations. Finally, we exemplified the materials design procedure by an example.

Acknowledgments

The authors thank the Chinese National Science Foundation (Nos. 21073118, 21376147), the Innovation Program of Shanghai Municipal Education Commission (13ZZ078), and the 085 Knowledge Innovation Program for the financial support, and acknowledge the

Laboratory for Microstructures, Shanghai University, for carrying out the structural characterization.

References

1. G. Ceder, Opportunities and challenges for first-principles materials design and applications to Li battery materials. *Mater. Res. Soc. Bull.* 2010, **35**(9), 693–701.
2. N. Agmon, The Grotthuss mechanism. *Chem. Phys. Lett.* 1995, **244**(5-6), 456–462.
3. D. Marx, Proton transfer 200 years after von Grotthuss: Insights from ab initio simulations. *ChemPhysChem* 2006, **7**(9), 1848–1870.
4. S. J. Paddison and J. A. Elliott, Molecular modeling of the short-side-chain perfluorosulfonic acid membrane. *J. Phys. Chem. A* 2005, **109**(33), 7583–7593.
5. D. Zhang and L. Yan, Probing the acid–base equilibrium in acid-benzimidazole complexes by ^{1}H NMR spectra and density functional theory calculations. *J. Phys. Chem. B* 2010, **114**(38), 12234–12241.
6. L. Yan, Q. Feng, L. Xie, and D. Zhang, About the choice of protogenic group for polymer electrolyte membrane: Alkyl or aryl phosphonic acid? *Solid State Ionics* 2011, **190**(1), 8–17.
7. L. Xie, H. Liu, S. Han, B. Yue, and L. Yan, Hydrogen bond and proton transport in acid–base complexes and amphoteric molecules by density functional theory calculations and ^{1}H and ^{31}P nuclear magnetic resonance spectroscopy. *J. Phys. Chem. B* 2013, **117**(50), 16345–16355.
8. B. Yue, L. Yan, S. Han, and L. Xie, Proton transport pathways in acid–base complex consisting of phosphonic acid group and 1,2,3-triazolyl group. *J. Phys. Chem. B* **2013**, 117(26), 7941–7949
9. S. J. Paddison, K. D. Kreuer, and J. Maier, About the choice of the protogenic group in polymer electrolyte membranes: Ab initio modelling of sulfonic acid, phosphonic acid, and imidazole functionalized alkanes. *Phys. Chem. Chem. Phys.* 2006, **8**(39), 4530–4542.
10. C. Wang and S. J. Paddison, Proton transfer in functionalized phosphonic acid molecules. *Phys. Chem. Chem. Phys.* 2010, **12**(4), 970–981.
11. L. Vilciauskas, S. J. Paddison, and K. D. Kreuer, Ab initio modeling of proton transfer in phosphoric acid clusters. *J. Phys. Chem. A* 2009, **113**(32), 9193–9201.

12. R. P. Pereira, M. I. Felisberti, and A. M. Rocco, Intermolecular interactions and formation of the hydration sphere in phosphonic acid model systems as an approach to the description of vinyl phosphonic acid based polymers. *Polymer* 2006, **47**(4), 1414–1422.

13. V. A. Glezakou, M. Dupuis, and C. J. Mundy, Acid/base equilibria in clusters and their role in proton exchange membranes: Computational insight. *Phys. Chem. Chem. Phys.* 2007, **9**(43), 5752–5760.

14. J. W. Blanchard, T. L. Groy, J. L. Yarger, and G. P. Holland, Investigating hydrogen-bonded phosphonic acids with proton ultrafast MAS NMR and DFT calculations. *J. Phys. Chem. C* 2012, **116**(35), 18824–18830.

15. Z. Zhou, R. Liu, J. Wang, S. Li, M. Liu, and J.-L. Brédas, Intra- and intermolecular proton transfer in $1H(2H)$-1,2,3-triazole based systems. *J. Phys. Chem. A* 2006, **110**(7), 2322–2324.

16. C. Nagamani, C. Versek, M. Thorn, M. T. Tuominen, and S. Thayumanavan, Proton conduction in $1H$-1,2,3-triazole polymers: Imidazole-like or pyrazole-like? *J. Polym. Sci., Part A: Polym. Chem.* 2010, **48**(9), 1851–1858.

17. C. Nagamani, U. Viswanathan, C. Versek, M. T. Tuominen, S. M. Auerbach, and S. Thayumanavan, Importance of dynamic hydrogen bonds and reorientation barriers in proton transport. *Chem. Commun.* 2011, **47**, 6638–6640.

18. U. Viswanathan, D. Basak, D. Venkataraman, J. T. Fermann, and S. M. Auerbach, Modeling energy landscapes of proton motion in nonaqueous, tethered proton wires. *J. Phys. Chem. A* 2011, **115**, 5423–5434.

19. G. F. Mangiatordi, J. Hermet, and C. Adamo, Modeling proton transfer in imidazole-like dimers: A density functional theory study. *J. Phys. Chem. A* 2011, **115**(12), 2627–2634.

20. G. F. Mangiatordi, V. Butera, N. Russo, D. Laage, and C. Adamo, Charge transport in poly-imidazole membranes: A fresh appraisal of the Grotthuss mechanism. *Phys. Chem. Chem. Phys.* 2012, **31**, 10910–10918.

21. G. F. Mangiatordi, D. Laage, and C. Adamo, Backbone effects on the charge transport in poly-imidazole membranes: A theoretical study. *J. Mater. Chem. A* 2013, **1**(26), 7751–7759.

22. C. C. Wu, C. Chaudhuri, J. C. Jiang, Y. T. Lee, and H. C. Chang, Structural isomerism and competitive proton solvation between methanol and water in $H^+(CH_3OH)_m(H_2O)_n$, $m+n=4$. *J. Phys. Chem. A* 2004, **108**(15), 2859–2866.

23. J. C. Jiang, C. Chaudhuri, Y. T. Lee, and H. C. Chang, Hydrogen bond rearrangements and interconversions of $H^+(CH_3OH)_4H_2O$ cluster isomers. *J. Phys. Chem. A* 2002, **106**(45), 10937–10944.

24. M. V. Mane, A. Venkatnathan, K. Ghatak, and K. Vanka, Exploring the potential of doped zero-dimensional cages for proton transfer in cuel cells: A computational study. *J. Phys. Chem. B* 2012, **116**(32), 9803–9811.

25. J. A. Elliott and S. J. Paddison, Modelling of morphology and proton transport in PFSA membranes. *Phys. Chem. Chem. Phys.* **2007**, **9**(21), 2602–2618.

26. I. H. Hristov, S. J. Paddison, and R. Paul, Molecular modeling of proton transport in the short-side-chain perfluorosulfonic acid ionomer. *J. Phys. Chem. B* 2008, **112**(10), 2937–2949.

27. J. A. Harvey, D. Basak, D. Venkataraman, and S. M. Auerbach, Simulating hydrogen-bond clustering and phase behaviour of imidazole oligomers. *Mol. Phys.* 2012, **110**(9-10), 1–10.

28. S. Zhu, L. Yan, D. Zhang, and Q. Feng, Molecular dynamics simulation of microscopic structure and hydrogen bond network of the pristine and phosphoric acid doped polybenzimidazole. *Polymer* 2011, **52**(3), 881–892.

29. T. Li, A. Wlaschin, and P. B. Balbuena, Theoretical studies of proton transfer in water and model polymer electrolyte systems. *Ind. Eng, Chem. Res.* 2001, **40**(22), 4789–4800.

30. E. Tsuchida, Ab initio molecular-dynamics simulation of concentrated phosphoric acid. *J. Phys. Soc. Jpn.* 2006, **75**(5), 054801–5.

31. D. W. M. Hofmann, L. N. Kuleshova, and B. D'Aguanno, Theoretical simulations of proton conductivity: Basic principles for improving the proton conductor. *J. Power Sources* 2010, **195**(23), 7743–7750.

32. R. L. Hayes, S. J. Paddison, and M. E. Tuckerman, Proton transport in triflic acid pentahydrate studied via ab initio path integral molecular dynamics. *J. Phys. Chem. A* 2011, **115**(23), 6112–6124.

33. G. A. Ludueña, T. D. Kühne, and D. Sebastiani, Mixed Grotthuss and vehicle transport mechanism in proton conducting polymers from ab initio molecular dynamics simulations. *Chem. Mater.* 2011, **23**(6), 1424–1429.

34. Q. F. Wang, N. S. Suraweera, D. J. Keffer, S. X. Deng, and J. Mays, Atomistic and coarse-grained molecular dynamics simulation of a cross-linked sulfonated poly(1,3-cyclohexadiene)-based proton exchange membrane. *Macromolecules* 2012, **45**(16), 6669–6685.

35. X. Ji, L. Yan, S. Zhu, L. Zhang, and W. Lu, Methanol distribution and electroosmotic drag in hydrated poly(perfluorosulfonic) acid membrane. *J. Phys. Chem. B* 2008, **112**(49), 15616–15627.
36. L. Yan, X. Ji, and W. Lu, Molecular dynamics simulations of electroosmosis in perfluorosulfonic acid polymer. *J. Phys. Chem. B* 2008, **112**(18), 5602–5610.
37. L. Yan, C. Shao, and X. Ji, Evaluation of electroosmotic drag coefficient of water in hydrated sodium perfluorosulfonate electrolyte polymer. *J. Comput. Chem.* 2009, **30**(9), 1361–1370.
38. C. Shao, L. Yan, X. Ji, and S. Zhu, Intermolecular momentum transfer in poly(perfluorosulfonic acid) membrane hydrated by aqueous solution of methanol: A molecular dynamics simulation study. *J. Chem. Phys.* 2009, **131**(22), 4901–4908.
39. P. Tölle, C. Köhler, R. Marschall, M. Sharifi, M. Wark, and T. Frauenheim, Proton transport in functionalised additives for PEM fuel cells: Contributions from atomistic simulations. *Chem. Soc. Rev.* 2012, **41**(15), 5143–5159.
40. K.-D. Kreuer, S. J. Paddison, E. Spohr, and M. Schuster, Transport in proton conductors for fuel-cell applications: Simulations, elementary reactions, and phenomenology. *Chem. Rev.* 2004, **104**(10), 4637–4678.
41. S. P. Brown and H. W. Spiess, Advanced solid-state NMR methods for the elucidation of structure and dynamics of molecular, macromolecular, and supramolecular systems. *Chem. Rev.* 2001, **101**(12), 4125–4156.
42. M. L. DiVona, E. Sgreccia, and S. Tosto, *Diffusion in solid proton conductors: Theoretical aspects and nuclear magnetic resonance analysis,* in *Solid State Proton Conductors* 2012, John Wiley & Sons, Ltd. pp. 25–70.
43. S. Suarez and S. Greenbaum, Nuclear magnetic resonance of polymer electrolyte membrane fuel cells. *Chem. Rec.* 2010, **10**(6), 377–393.
44. H. W. Spiess, Interplay of structure and dynamics in macromolecular and supramolecular systems. *Macromolecules* 2010, **43**(13), 5479–5491.
45. A. Schechter, R. F. Savinell, J. S. Wainright, and D. Ray, ^{1}H and ^{31}P NMR study of phosphoric acid-doped polybenzimidazole under controlled water activity. *J. Electrochem. Soc.* 2009, **156**(2), B283–B290.
46. S. Wasmus, A. Valeriu, G. D. Mateescu, D. A. Tryk, and R. F. Savinell, Characterization of H_3PO_4-equilibrated Nafion® 117 membranes using ^{1}H and ^{31}P NMR spectroscopy. *Solid State Ionics* 1995, **80**(1-2), 87–92.
47. S. Toppet, G. Wouters, and G. Smets, ^{13}C NMR study of poly(4-vinyltriazole). Direct observation of the three tautomers. *J. Polym. Sci.: Polym. Lett. Ed.* 1976, **14**(7), 389–394.

48. G. R. Goward, M. F. H. Schuster, D. Sebastiani, I. Schnell, and H. W. Spiess, High-resolution solid-state NMR studies of imidazole-based proton conductors: Structure motifs and chemical exchange from ^{1}H NMR. *J. Phys. Chem. B* 2002, 106(36), 9322–9334.

49. S. U. Çelik, Ü. Akbey, R. Graf, A. Bozkurt, and H. W. Spiess, Anhydrous proton-conducting properties of triazole-phosphonic acid copolymers: A combined study with MAS NMR. *Phys. Chem. Chem. Phys.* 2008, **10**(39), 6058–6066.

50. Ü. Akbey, R. Graf, P. P. Chu, and H. W. Spiess, Anhydrous poly(2,5-benzimidazole)-poly(vinylphosphonic acid) acid–base polymer blends: A detailed solid-state NMR investigation. *Aust. J. Chem.* 2009, **62**(8), 848–856.

51. Ü. Akbey, R. Graf, Y. G. Peng, P. P. Chu, and H. W. Spiess, Solid-state NMR investigations of anhydrous proton-conducting acid–base poly(acrylic acid)-poly(4-vinyl pyridine) polymer blend system: A study of hydrogen bonding and proton conduction. *J. Polym. Sci., Part B: Polym. Phys.* 2009, **47**(2), 138–155.

52. Ü. Akbey, S. Granados-Focil, E. B. Coughlin, R. Graf, and H. W. Spiess, ^{1}H solid-state NMR investigation of structure and dynamics of anhydrous proton conducting triazole-functionalized siloxane polymers. *J. Phys. Chem. B* 2009, **113**(27), 9151–9160.

53. R. Frantz, J.-O. Durand, F. Carré, G. F. Lanneau, J. L. Bideau, B. Alonso, and D. Massiot, Synthesis and solid-state NMR studies of *p*-vinylbenzylphosphonic acid. *Chem. Eur. J.* 2003, **9**(3), 770–775.

54. Y. J. Lee, B. Bingöl, T. Murakhtina, D. Sebastiani, W. H. Meyer, G. Wegner, and H. W. Spiess, High-resolution solid-state NMR studies of poly(vinyl phosphonic acid) proton-conducting polymer: Molecular structure and proton dynamics. *J. Phys. Chem. B* 2007, **111**(33), 9711–9721.

55. G. Brunklaus, S. Schauff, D. Markova, M. Klapper, K. Müllen, and H.-W. Spiess, Proton mobilities in phosphonic acid–based proton exchange membranes probed by ^{1}H and ^{2}H solid-state NMR spectroscopy. *J. Phys. Chem. B* **2009**, **113**(19), 6674–6681.

56. I. Fischbach, H. W. Spiess, K. Saalwachter, and G. R. Goward, Solid state NMR spectroscopic investigations of model compounds for imidazole-based proton conductors. *J. Phys. Chem. B* 2004, **108**(48), 18500–18508.

57. S. R. Benhabbour, R. P. Chapman, G. Scharfenberger, W. H. Meyer, and G. R. Goward, Study of imidazole-based proton-conducting composite materials using solid-state NMR. *Chem. Mater.* 2005, **17**(6), 1605–1612.

58. H. A. Every, L. D. Ionescu, M. P. d. Heer, Y. Álvarez-Gallego, and G. J. M. Janssen, NMR studies of proton transport in anhydrous polymer electrolytes for high temperature fuel cells. *Fuel Cells* 2008, **8**(3-4), 254–261.
59. J. W. Traer, E. Montoneri, A. Samoson, J. Past, T. Tuherm, and G. R. Goward, Unraveling the complex hydrogen bonding of a dual-functionality proton conductor using ultrafast magic angle spinning NMR. *Chem. Mater.* 2006, **18**(20), 4747–4754.
60. S. Selvasekarapandian, G. Hirankumar, J. Kawamura, N. Kuwata, and T. Hattori, ^{1}H solid state NMR studies on the proton conducting polymer electrolytes. *Mater. Lett.* 2005, **59**(22), 2741–2745.
61. C. E. Hughes, S. Haufe, B. Angerstein, R. Kalim, U. Mahr, A. Reiche, and M. Baldus, Probing structure and dynamics in *p*[2,2'-(*m*-phenylene)-5,5'-bibenzimidazole] fuel cells with magic-angle spinning NMR. *J. Phys. Chem. B* 2004, **108**(36), 13626–13631.
62. G. Ye, N. Janzen, and G. R. Goward, Solid-state NMR study of two classic proton conducting polymers: Nafion and sulfonated poly(ether ether ketone)s. *Macromolecules* **2006**, **39**(9), 3283–3290.
63. Z. J. Li, Y. Abramov, J. Bordner, J. Leonard, A. Medek, and A. V. Trask, Solid-state acid–base interactions in complexes of heterocyclic bases with dicarboxylic acids: Crystallography, hydrogen bond analysis, and ^{15}N NMR spectroscopy. *J. Am. Chem. Soc.* 2006, **128**(25), 8199–8210.
64. Y. J. Lee, T. Murakhtina, D. Sebastiani, and H. W. Spiess, ^{2}H solid-state NMR of mobile protons: It is not always the simple way. *J. Am. Chem. Soc.* 2007, **129**(41), 12406–12407.
65. T. A. Zawodzinski, M. Neeman, L. O. Sillerud, and S. Gottesfeld, Determination of water diffusion coefficients in perfluorosulfonate ionomeric membranes. *J. Phys. Chem.* 1991, **95**(15), 6040–6044.
66. H. A. Every, M. A. Hickner, J. E. McGrath, and J. T. A. Zawodzinski, An NMR study of methanol diffusion in polymer electrolyte fuel cell membranes. *J. Membr. Sci.* 2005, **250**(1-2), 183–188.
67. K. Kidena, T. Ohkubo, N. Takimoto, and A. Ohira, PFG-NMR approach to determining the water transport mechanism in polymer electrolyte membranes conditioned at different temperatures. *Eur. Polym. J.* 2010, **46**(3), 450–455.
68. X. M. Ren, T. E. Springer, T. A. Zawodzinski, and S. Gottesfeld, Methanol transport through nafion membranes - electroosmotic drag effects on potential step measurements. *J. Electrochem. Soc.* 2000, **147**(2), 466–474.

69. I. Nicotera, A. Khalfan, G. Goenaga, T. Zhang, A. Bocarsly, and S. Greenbaum, NMR investigation of water and methanol mobility in nanocomposite fuel cell membranes. *Ionics* 2008, **14**(3), 243–253.

70. A. S. Aricò, V. Baglio, V. Antonucci, I. Nicotera, C. Oliviero, L. Coppola, and P. L. Antonucci, An NMR and SAXS investigation of DMFC composite recast NAFION membranes containing ceramic fillers. *J. Membr. Sci.* 2006, **270**(1-2), 221–227.

71. J. R. P. Jayakody, A. Khalfan, E. S. Mananga, S. G. Greenbaum, T. D. Dang, and R. Mantz, NMR investigation of water and methanol transport in sulfonated polyareylenethioethersulfones for fuel cell applications. *J. Power Sources* 2006, **156**(2) 195–199.

72. V. I. Volkov and E. V. Volkov, Ionic and molecular self-diffusion in ion-exchange materials for fuel energetics studied by pulsed field gradient NMR. *Appl. Magn. Reson.* 2005, **29**(3), 495–502.

73. Y. F. Huang, L. C. Chuang, A. M. Kannan, and C. W. Lin, Proton-conducting membranes with high selectivity from cross-linked poly(vinyl alcohol) and poly(vinyl pyrrolidone) for direct methanol fuel cell applications. *J. Power Sources* 2009, **186**(1), 22–28.

74. C. W. Lin, S. S. Liang, S. W. Chen, and J. T. Lai, Sorption and transport properties of 2-acrylamido-2-methyl-1-propanesulfonic acid-grafted bacterial cellulose membranes for fuel cell application. *J. Power Sources* 2013, **232**(0), 297–305.

75. S. Suarez, N. K. Kodiweera, P. Stallworth, S. Yu, S. G. Greenbaum, and B. C. Benicewicz, Multinuclear NMR study of the effect of acid concentration on ion transport in phosphoric acid doped poly(benzimidazole) membranes. *J. Phys. Chem. B* 2012, **116**(41), 12545–12551.

76. J. R. P. Jayakody, S. H. Chung, L. Durantino, H. Zhang, L. Xiao, B. C. Benicewicz, and S. G. Greenbaum, NMR studies of mass transport in high-acid-content fuel cell membranes based on phosphoric acid and polybenzimidazole. *J. Electrochem. Soc.* 2007, **154**(2), B242–B246.

77. I. Nicotera, K. Angjeli, L. Coppola, A. S. Aricò, and V. Baglio, NMR and electrochemical investigation of the transport properties of methanol and water in Nafion and clay-nanocomposites membranes for DMFCs. *Membranes* 2012, **2**(2), 325–345.

78. V. I. Volkov, Y. M. Popkov, S. F. Timashev, D. G. Bessarabov, R. D. Sanderson, and Z. Twardowski, Self-diffusion of water and fluorine ions in anion-exchange polymeric materials (membranes and resin) as determined by pulsed-field gradient nuclear magnetic resonance spectroscopy. *J. Membr. Sci.* 2000, **180**(1), 1–13.

79. T. M. Alam and M. R. Hibbs, Characterization of heterogeneous solvent diffusion environments in anion exchange membranes. *Macromolecules* 2014, **47**(3), 1073–1084.

80. M. Schuster, T. Rager, A. Noda, K. D. Kreuer, and J. Maier, About the choice of the protogenic group in PEM separator materials for intermediate temperature, low humidity operation: A critical comparison of sulfonic acid, phosphonic acid and imidazole functionalized model compounds. *Fuel Cells* 2005, **5**(3), 355–365.

81. V. I. Volkov and A. A. Marinin, NMR methods for studying ion and molecular transport in polymer electrolytes. *Russ. Chem. Rev.* 2013, **82**(3), 248.

82. M. Ise, K. D. Kreuer, and J. Maier, Electroosmotic drag in polymer electrolyte membranes: An electrophoretic NMR study. *Solid State Ionics* 1999, **125**(1-4), 213–223.

83. J. Kawamura, K. Hattori, T. Hongo, R. Asayama, N. Kuwata, T. Hattori, and J. Mizusaki, Microscopic states of water and methanol in Nafion membrane observed by NMR micro imaging. *Solid State Ionics* 2005, **176**(31-34), 2451–2456.

84. M. J. Frisch, G. W. Trucks, H. B. Schlegel, G. E. Scuseria, M. A. Robb, J. R. Cheeseman, J. A. Montgomery, T. Vreven Jr, K. N. Kudin, J. C. Burant, et al., *GAUSSIAN 03, Revision C.2*, 2003, Gaussian, Inc.: Wallingford, CT.

85. M.-H. Hao, Theoretical calculation of hydrogen-bonding strength for drug molecules. *J. Chem. Theory Comput.* 2006, **2**(3), 863–872.

86. M. Nocker, S. Handschuh, C. Tautermann, and K. R. Liedl, Theoretical prediction of hydrogen bond strength for use in molecular modeling. *J. Chem. Inf. Model.* 2009, **49**(9), 2067–2076.

87. S. Boys and F. Bernardi, The calculation of small molecular interactions by differences of separate total energies - some procedures with reduced errors. *Mol. Phys.* 1970, **19**(4), 553–566.

88. S. Simon, M. Duran, and J. J. Dannenberg, How does basis set superposition error change the potential surfaces for hydrogen-bonded dimers? *J. Chem. Phys.* 1996, **105**(24), 11024–11031.

89. L. Yan and L. Xie, Molecular dynamics simulations of proton transport in proton exchange membranes based on acid–base complexes, in *Molecular Interaction*, A. Meghea, Editor 2012, InTech: Rijeka, Croatia. pp. 335–360.

90. S. L. Mayo, B. D. Olafson, and W. A. Goddard III, DREIDING: A generic force field for molecular simulations. *J. Phys. Chem.* 1990, **94**(26), 8897–8909.

91. W. Smith, M. Leslie, and T. R. Forester, *Computer code DL_POLY_2.14*. 2003: CCLRC, Daresbury Laboratory, Daresbury, England.
92. S. Nosé, A unified formulation of the constant temperature molecular dynamics methods. *J. Chem. Phys.* 1984, **81**(1), 511–519.
93. W. G. Hoover, Canonical dynamics: Equilibrium phase-space distributions. *Phys. Rev. A* 1985, **31**(3), 1695–1697.
94. L. Verlet, Computer "Experiments" on classical fluids. I. Thermodynamical properties of Lennard-Jones molecules. *Phys. Rev.* 1967, **159**(1), 98–103.
95. R. Dhani, A. Avinash, S. K. Salenaagina, M. V. S. Teja, P. Masthanaiah, P. R. Rathnam, and V. C. Silpa, Indole: The molecule of diverse pharmacological activities. *J. Chem. Pharm. Res.* 2011, **3**(5), 519–523.

Chapter 6

Nanomaterials for Oxygen Reduction Reaction (ORR)

R. K. Singh,[a] F. G. S. Wasim,[a,b] and M. Neergat[a]

[a]*Department of Energy Science and Engineering and*
[b]*IITB-Monash Research Academy, Indian Institute of Technology Bombay (IITB), Powai, Mumbai, Maharashtra 400076, India*
nmanoj@iitb.ac.in

Nanomaterials have tremendous implications in various disciplines of science and technology. The physical, chemical, and electrochemical properties of these materials can be tuned by controlling their size, shape, surface structure, and compositions (both surface and bulk) though it is a difficult task at a nanoscale dimension. In this chapter, nanomaterials, both precious and non-precious metal-based, relevant to oxygen reduction reaction (ORR) are discussed after a brief introduction to fuel cells and the mechanism of ORR. Methods for estimating the ORR product selectivity (H_2O/H_2O_2) and the electrochemical surface area (ESA) are introduced followed by the descriptor of activity. The evolution of electrocatalyst from the unsupported to core–shell structure is presented. These include precious metal black, carbon-supported alloys of precious metals (Pt and Pd) with transition metals, shape-controlled, and core–shell catalysts besides the non-precious metal catalysts. The main focus is

Nanomaterials for Direct Alcohol Fuel Cell
Edited by Yixuan Wang

ISBN 978-981-4669-00-9 (Hardcover), 978-981-4669-01-6 (eBook)
www.panstanford.com

on the chemical synthesis of the above-mentioned materials, surface cleaning, and their activity toward ORR. The origin of the ORR activity improvement is summarized. The chapter is concluded with emphasis on the need for further research efforts towards enhancing the electrocatalytic activity, long-term stability, and reduction of the cost. Overall, the investigations on ORR significantly contribute to the better understanding of the electrochemistry of nanomaterials.

6.1 Introduction

Fuel cell is an electrochemical energy conversion device that directly converts chemical energy of the fuel into electrical energy. The electrochemical energy conversion process is considered to be very efficient since it does not involve Carnot efficiency limitations. Any fuel cell device consists of an anode, where the fuel gets electrochemically oxidized; a cathode, where the oxidant, air or oxygen, gets reduced; and an electrolyte that transports the ions either from the anode to the cathode (cationic electrolytes) or from the cathode to the anode (anionic electrolytes). Depending on the type of the ions conducted by the electrolyte medium, fuel cells operate at low, medium, or high temperature ranges. The low-temperature fuel cells use proton-conducting polymer electrolyte membranes (PEMs) or cation exchange membranes as the ion transport medium. These ion-conducting membranes also function as a separator between the anode and the cathode avoiding any short circuit and preventing the direct contact between the fuel and the oxidant. Among various fuel cells available, low-temperature fuel cells have attracted significant research attention due to their possible applications in mobile and transport sectors. Thus, significant research efforts have been expended on the development of low-temperature fuel cells.

Low-temperature fuel cells typically use hydrogen or lower aliphatic alcohols such as methanol and ethanol as the fuel. The electrochemical oxidation of hydrogen, the most commonly used fuel, is a kinetically facile process; therefore, the polarization losses are negligible at the anode and it requires only minute quantity of the Pt catalyst. Thus, anode catalyst development is significantly limited in the last one decade. On the other hand, alcohol fuels such as ethanol and methanol are difficult to oxidize and there are significant polarization losses arising from the fuel oxidation. Recently, sodium borohydride is also used as a fuel in the low-temperature fuel cells. The use of these fuels simplifies the overall fuel cell system design by reducing the complexity of fuel reforming.

The complex four-electron reduction of oxygen to water, the reaction occurring at the cathode side in fuel cells, has attracted significant research interest. This is mostly due to the very high overpotential losses—one of the major contributors to the fuel cell performance loss. Thus, the primary research focus has been on the development of materials for the oxygen reduction reaction (ORR). In this chapter, various nanomaterial-based approaches that are reported in the literature for the electrocatalysis of ORR are discussed.

6.2 Methodology and Descriptor of Activity

Before analyzing the performance and characteristics of each catalyst, the descriptor of activity deserves a brief discussion. The straightforward way of comparing the quality of the cathode catalysts is to test their performance against a standard anode in a fuel cell configuration by fabricating membrane electrode assemblies (MEAs); in an operating fuel cell, it is hard to device a reference electrode. However, the fuel cell operating conditions (temperature, pressure, oxygen flow-rate, and humidity) and the MEA fabrication procedures adopted by various research groups are different. Therefore, it is complex to estimate the true potential of the catalysts and to compare their activities and activity improvement. Characterization of the catalyst in an operating fuel cell mode is often expensive and an alternative is the rotating disk electrode/rotating ring disk electrode (RDE/RRDE) configuration [1]. This widely used technique requires only a minute quantity of the catalyst (typically $\sim$15 μg cm^{-2}, two orders of magnitude less catalyst loading as compared to that in fuel cell MEAs) [2]. It is a single-electrode study in a three-electrode configuration without involving the membrane electrolyte and the anode catalyst layer.

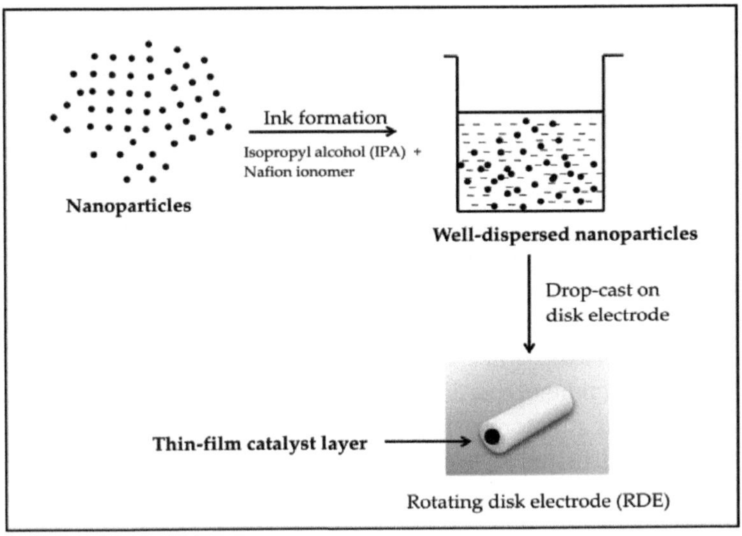

Figure 6.1 Scheme for the formation of thin-film catalyst layer on RDE.

Thus, if the film formation is correct and reproducible, the data is very much reliable. Typically, the polarization data are recorded in dilute acidic (0.1 M $HClO_4$ or 0.5 M H_2SO_4) and alkaline (0.1 M NaOH or KOH) electrolytes [3, 4]. Scheme for the formation of thin-film catalyst layer is shown in Fig. 6.1.

For comparison of the performance of the catalysts, one can either use the half-wave potential of ORR or mass and specific activities. While it is fair to compare the activity in terms of half-wave potential of ORR on single crystal electrodes with roughness factor (electrochemical surface area of the catalyst/electrode geometrical area) of ~1, it is more appropriate to compare specific activity with polycrystalline catalysts of different roughness factors. The mass activity is obtained by normalizing the mass transfer, and IR-corrected kinetic current with catalyst loading and the specific activity is obtained by normalizing the kinetic current with the area of the catalyst on the disk electrode. Thus, the estimation of the electrochemical surface area (ESA) of the catalyst is very important (Table A1 shows the mass and specific activities of various Pt and Pd-based catalysts). ESA in-turn is calculated from the hydrogen under potential deposition (H_{upd}) region of the voltammogram, Cu stripping, and CO stripping voltammograms. These methods are quite reliable with precious metal catalysts in acidic electrolytes. A brief discussion of the ESA calculation is given below.

6.2.1 Calculation of the ESA of Pt/C in Acidic Media

The ESA of precious metal catalysts is derived from the H_{upd} region of the voltammogram or from Cu/CO stripping voltammogram. The CV in acidic medium consists of an H_{upd} region (0.05–0.4 V), double-layer region (0.4–0.8 V), and oxide formation region (>~0.8 V). It is imperative to subtract the background double-layer charge while estimating the ESA using any of the afore-mentioned methods. After baseline correction, the charge under the hydrogen adsorption (H_{ads}) or desorption (H_{des}) peak or Cu/CO stripping peak is integrated to calculate the ESA. Typical schematic of the stripping voltammograms along with the baseline voltammogram is shown in Fig. 6.2. A conversion factor of 210 μC cm^{-2} (for 1-e$^-$ process) is used with H_{upd} and 410 μC cm^{-2} (for 2-e$^-$ process) with Cu/CO stripping methods [5, 6].

$$\text{Electrochemical surface area (ESA)} = \frac{\text{Charge (C)}}{\text{Conversion factor}\left(\mu\text{C cm}^{-2}\right) \times \text{Loading}\left(\mu\text{g}\right)} \tag{6.1}$$

6.3 Reaction Mechanism and Thermodynamics

6.3.1 Pathways in Acidic Medium

The mechanism of four-electron reduction of oxygen in acidic medium suggested by Wroblowa et al. is shown below [7].

$$O_2 + 4H^+ + 4e^- \rightarrow 2H_2O \tag{6.2}$$

The mechanism suggests that H_2O_2 is formed parallel to the reduction of O_2 to water. In the mechanism, k_i represents the rate constant, and indices *b*, *, and *a*, designate the bulk of the electrolyte, vicinity of the electrode surface, and adsorbed species, respectively.

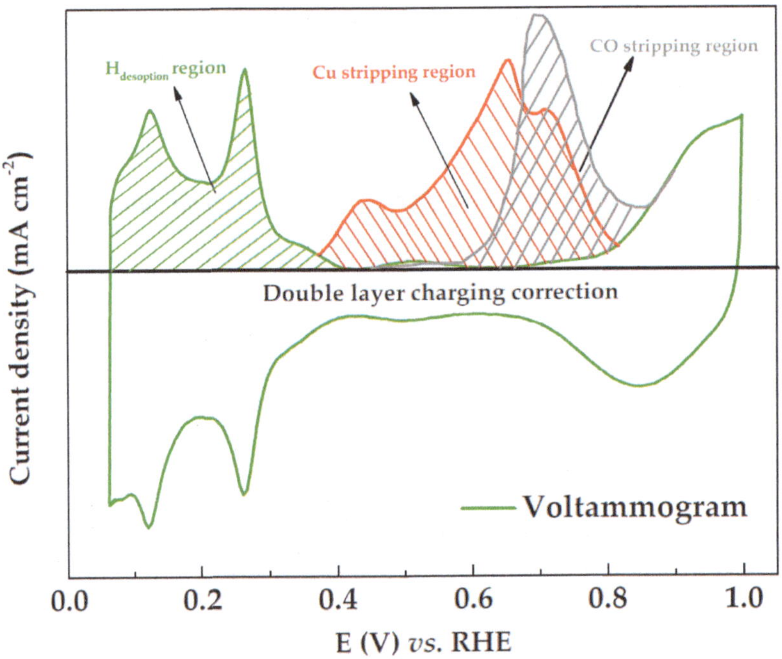

Figure 6.2 Schematic of the voltammogram of polycrystalline Pt recorded in an argon-saturated 0.1 M H_2SO_4 solution (green solid line), baseline-corrected Cu stripping (red solid line), and CO stripping voltammograms (grey solid line). The shaded area is used for the ESA calculation from H_{upd}, Cu stripping, and CO stripping.

6.3.2 Detection of Peroxide

RRDE, a combination of the RDE with ring electrode, can be used to estimate the quantity of H_2O_2 with the following formula:

$$X_{H_2O_2} = 2\frac{I_R}{N} \Big/ \left(I_D + \frac{I_R}{N}\right) \tag{6.3}$$

Here, $X_{H_2O_2}$ is the fraction of the peroxide, I_R the ring current, I_D the disk current, and N the collection efficiency of RRDE [1].

This method is based on the oxidation of H_2O_2, a by-product of O_2 reduction reaction (see the mechanism discussed above), on the ring electrode at a potential of ~1.2 V. Peroxide generation features can provide significant insight into the ORR mechanism.

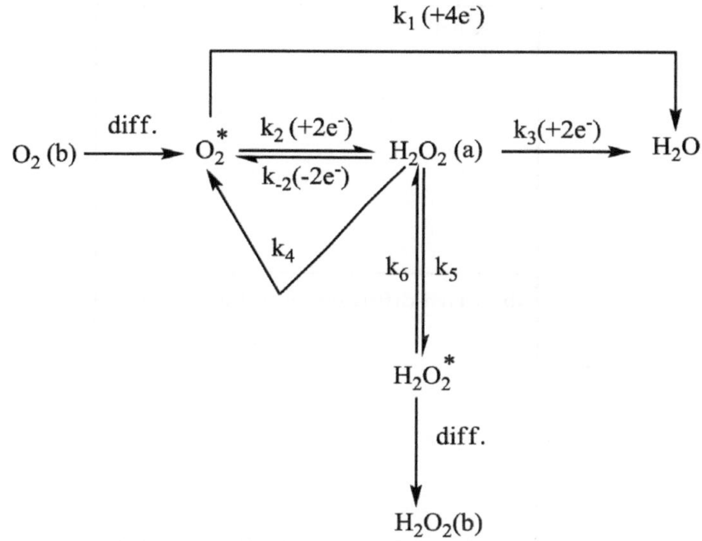

Scheme 6.1 ORR pathway as suggested by Wroblowa et al. [7]. Reprinted with permission, copyright © 1976 Elsevier.

6.3.3 Electrochemical Impedance Spectroscopy (EIS)

The electrochemical impedance spectroscopy (EIS) can be highly effective in characterizing the nanoparticles for fuel cell applications. For example, the schematic representation of the EI spectra of ORR on Pt/C (20 wt%) is shown in Fig. 6.3. In the kinetic region, the spectra consist of an arc or semicircle. The radius of the semicircle decreases with overpotential and it represents the polarization resistance; it is dominated by the charge transfer resistance (Fig. 6.3a). From the charge transfer resistance at high potential, the exchange current density can be estimated. In the mixed diffusion-controlled region, the impedance spectrum consists of an arc or 45° straight line in the high-frequency (HF) region in addition to the two semicircles corresponding to the charge transfer and mass transfer resistances (see Fig. 6.3b). A 45° straight line at the HF region originates from the diffusion limitation of the protons through the catalyst layer [8–10]. From the low-frequency semicircle, the mass transfer resistance can be estimated. The EIS features depend on the metal loading, carbon-support loading, and ionomer content

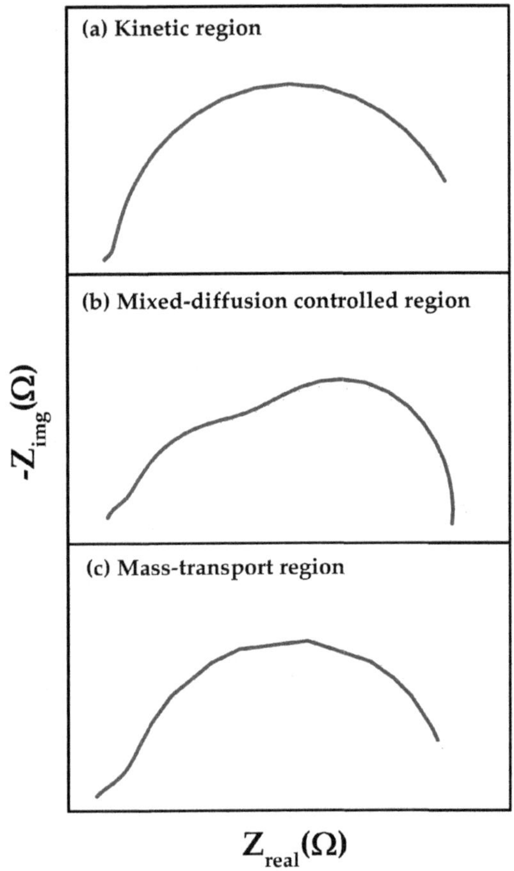

Figure 6.3 Schematic representation of the EI spectra of ORR on Pt/C: (a) kinetic region, (b) mixed diffusion-controlled region, and (c) mass transport region.

in the catalyst layer. In the lower-potential region, the charge transfer resistance further decreases and hence a single semicircle representing both the charge transfer and mass transfer resistances appears with major contribution from the latter (Fig. 6.3c). Thus, impedance spectroscopy is a versatile tool to investigate the kinetic parameters of the electrochemical reaction. But caution is required in the analysis of the spectrum. EIS can be very useful for the characterization of non-precious metal catalysts (NPMCs) with low

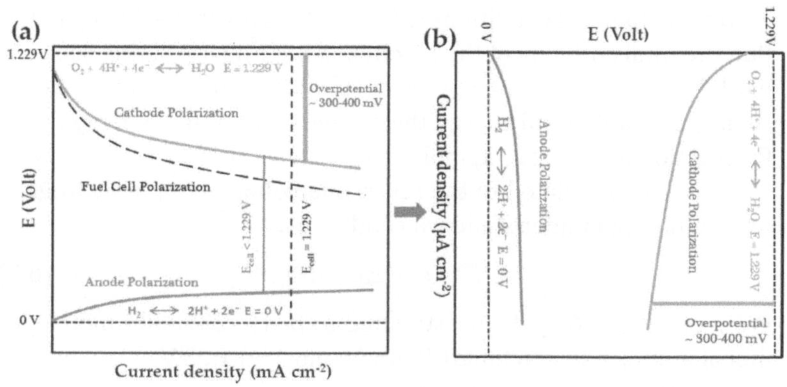

Figure 6.4 Schematic representations of (a) anode, cathode, and fuel cell polarizations and (b) single electrode system in an RDE configuration.

turn-over frequencies (ToFs) where the mass transfer limitations are very high. It can also be used to estimate the ESA of the catalyst and diffusion coefficient of the reacting species.

6.4 Fuel Cell and Single-Electrode (RDE) Polarizations

The fuel cell reaction consists of two half-electrochemical cell reactions. At the anode, fuel gets oxidized and at the cathode oxygen gets reduced. Schematic representations of both anode and cathode polarizations of a fuel cell and that obtained from an RDE are shown in Fig. 6.4. A paradigm shift in the research direction from fuel cell to single electrode is evident from the literature. This is mainly due to the realization that the component material properties need to be improved significantly.

6.4.1 Calculation of $\Delta G°$ and $E°$

The reversible cell voltage of a hydrogen–oxygen fuel cell can be calculated from the enthalpy change ($\Delta H°$) of reaction 6.4.

$$H_2 + 1/2O_2 \rightarrow H_2O \tag{6.4}$$

Under standard conditions (25°C, and 1 atm), it is -283 kJ mol^{-1}. But, the total enthalpy of the reaction is not utilized in the electrical work because of the entropy term (the losses to the atmosphere). Hence, the total useful energy, the change in Gibbs free energy (ΔG°) for reaction 6.4 is -237 kJ mol^{-1}.

Also, ΔG° is related to the reversible voltage (E°) of hydrogen–oxygen fuel cell under standard conditions.

$$E^\circ = -(\Delta G^\circ/\text{nF}) \tag{6.5}$$

From equation (6.5), E° is estimated to be ((−237000 J mol^{-1})/(2 mol of e$^-$mol^{-1} of reactants)(96485 C mol^{-1})) 1.23 V.

At any given temperature and pressure,

$$\Delta G = \Delta G^\circ - RT\ln K \tag{6.6}$$

where, K is the equilibrium constant.

Substituting Eq. 6.5 with Eq. 6.6,

$$E = E^\text{o} - \frac{RT}{nF}\ln K \tag{6.7}$$

where, E° is the standard-state reversible voltage and ΔG° is the standard state free-energy change of the reaction.

For any deviation from the equilibrium conditions, the cell voltage is given by Nernst equation.

$$E = E^\text{o} - \frac{RT}{2F}\ln\frac{a_{\text{H}_2\text{O}}}{a_{\text{H}_2}a_{\text{O}_2}^{1/2}} \tag{6.8}$$

Thus, the voltage of a H_2–O_2 fuel cell operating with pure hydrogen and oxygen at 1 atm pressure each at room temperature is 1.23 V, as shown below.

$$E = E^\text{o} - \frac{RT}{2F}\ln\frac{1}{p_{\text{H}_2}p_{\text{O}_2}^{1/2}}$$

$$= 1.23 - \frac{RT}{2F}\ln\frac{1}{(1)(1)} = 1.23\text{ V}$$

On replacing oxygen with air at 1 atm, since the oxygen content in air is only 21%, the E° decreases from the standard value to 1.214 V, as shown below.

$$E = 1.23 - \frac{RT}{2F}\ln\frac{1}{(1)(0.21)} = 1.214\text{ V}$$

But this theoretical cell voltage is never realized in a practical fuel cell. This happens because of the parallel formation of several non-ideal products along with the major product water as observed from the reaction mechanism. Moreover, at the open-circuit potential (OCP) close to 1.2 V, the Pt catalyst surface itself is susceptible to oxide formation. Thus, what is observed is a potential of $\sim$1 V, which is much lower than the theoretical OCP.

6.5 Recent Developments in Nanomaterials for ORR

Having discussed the fuel cell fundamentals and background on electrocatalysis, in the following, an overview of the recent developments in nanomaterials is presented. Nanomaterials that are used for the electrocatalysis of ORR can broadly be classified into precious and non-precious metal catalysts; among these, precious metal electrocatalysts are the most reported in the literature [11]. These include carbon-supported and unsupported catalysts, complex nanostructures, shape-controlled nanoparticles, shape-controlled alloy nanoparticles, and core–shell materials [1, 5, 12–14]. Non-precious metal catalysts include the first-generation chalcogenides, oxides, and macrocyclic compounds (phthalocyanines and porphirines) [15–17]. The new class of electrocatalysts includes composites of different non-precious metals like Fe, Co, and Ni with nitrogen and carbon black/carbon nanotubes (single- and multiwalled)/graphene [18–21]. In the following, we will discuss the recent developments in catalysts for ORR in low-temperature fuel cells. The flow chart shows the chronology of the Pt-based catalyst development (Fig. 6.5).

6.5.1 Precious Metal Catalysts

The penetration of fuel cells in the wider market is hindered by the sluggish ORR. Carbon-supported Pt (Pt/C), the state-of-the-art catalyst material, is expensive and the overpotential towards the ORR is very high at the cathode in low-temperature fuel cells (polymer electrolyte fuel cells (PEFCs) and direct methanol fuel

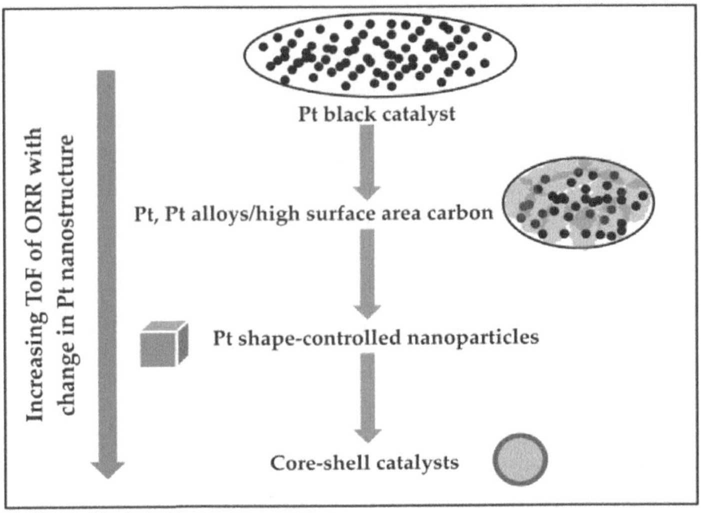

Figure 6.5 Evolution of the Pt-based catalysts.

cells (DMFCs)) [2, 22, 23]. Currently, in practical demonstration applications, such as the automotives, the Pt loading is 0.4 mg cm^{-2} or more on the cathode and the stability of the catalyst is still short of the 5000 hour durability target [11, 24]. Commercial Pt/C has specific activities in the range of 0.1–0.2 mA cm^{-2} Pt and mass activities of 0.1–0.12 A mg^{-1} Pt [11]. The performance target on Pt-based catalyst is an order of magnitude improvement in the specific activity. The perennial question is how to reduce the precious metal loading to <0.1 mg cm^{-2} without compromising the performance and durability.

Several approaches are reported to improve the activity and to reduce the cost of the catalyst. One of the initial approaches to reduce the platinum loading was to make the particle size as small as possible and the other was to produce supported Pt nanoparticles [25, 26]. The commercial Pt/C electrocatalyst, currently used in the fuel cells, has high catalyst dispersion (fraction of surface atoms to total number of atoms in a nanoparticle) of ~0.35 [27]. Carbon-supported colloidal Pt nanoparticles of diameter 2–5 nm are currently considered as the best catalysts for ORR. Obtaining a stable catalyst with lower particle size and increased dispersion is a difficult task and it is constrained by the limitations of the conventional synthesis procedures. Though the inherent activity of very small Pt nanoparticles (<2–5 nm) is not yet established, further miniaturization to sub-nanometer-scale might cause a decrease in catalytic activity. A definite answer to the size limit for the best activity remains elusive [26, 27]. Studies on electrocatalysis of ORR in acidic media with precious metal catalysts are primarily limited to Pt-based materials (Pt black catalysts, Pt/C and single crystals of Pt) and, to some extent, to Pd-based materials [4, 11, 14, 23, 28, 29].

The ORR kinetic limitations on these catalyst surfaces can be addressed by manipulating surface morphology (e.g., nano-architectural and shape-controlled materials based on both Pt and Pd) and by alloying the Pt-based material, i.e., by changing its electronic properties (e.g., Pt-based alloys and core–shell catalysts) [13, 14, 28, 30].

In order to improve the efficiency and to reduce the cost of fuel cell, an alternative material should be used as the catalyst. Alloying Pt with transition metals (TMs) is one of the methods reported for improving the activity of carbon-supported Pt catalysts (Table A1) [31–33]. The most widely reported alloy combination is Pt_3TM, (where TM is Cr, Fe, Co, Ni, Cu, Ti, and V) and among these, Pt_3Ni has shown the best activity, 90-fold more active than platinum [34]. Thus, alloying Pt with other elements has been shown to improve the kinetics of ORR significantly, while reducing the catalyst cost and the ORR activity improvement is well established with phosphoric acid fuel cells (PAFCs), PEFCs, and DMFCs [24, 32, 35].

Usually, these catalysts are synthesized by the high temperature heat-treatment of Pt/C with TM salts (e.g., nitrates, in inert or reducing atmosphere at temperature above ~600°C). Mostly, the enhancement in activity is explained on the basis of the following concepts. X-ray diffraction (XRD) patterns and transmission electron microscopy (TEM) of Pt suggest a face-centered cubic (FCC) structure with a lattice parameter of 3.93 Å. It has been observed that, while alloying, TM with relatively lower atomic radius enters the lattice of Pt through substitution. The enhanced ORR activity of the alloys has been attributed to the resulting lattice contraction that facilitates the facile dissociative-adsorption of molecular oxygen [36]. A few other hypotheses are also reported in the literature to explain the enhancement in the activity of Pt alloys relative to that of Pt.

Other investigators reported that the more easily oxidizable base metal dissolves or leaches from the bulk of the alloy in acid medium. The consequent roughening of the active alloy surface explains the ORR activity improvement, though the heat treatment causes an increase in crystallite size and a reduction in the ESA of the catalyst [37].

Mukerjee et al. reported the vital role of the electronic factors originating from the change in d-band vacancies of the precious metal component in the alloy catalyst [32]. Several studies demonstrated a reduction in Pt-Pt bond distance in Pt-TM alloys and consequent increase in d-band vacancy. XANES data reported in the literature show an upshift in the onset OH_{ads} formation potential with carbon-supported Pt-alloys as compared to that of Pt at 0.8 V. Thus, the activity improvement with Pt alloys (PtCr, PtCo and PtNi) is ascribed to the delay of OH_{ads} formation to higher potentials [38]. Formation of skeleton surfaces upon dissolution of the transition metal atoms in acidic solution and Pt skin-layer are also reported to affect the activity improvement [30].

Alloying platinum with other metals was found to improve the activity, but still, the reduction in platinum loading has to be achieved for widespread commercialization (Table A1). The stability and long-term durability of these alloy catalysts are also a challenge because of the leaching of transition metal in highly acidic PEMFC condition; the alloy component leached from the catalyst surface may cation-exchange with the H^+ ions of the conducting electrolyte leading to decrease in the ionic conductivity of the membrane electrolyte. But even after long-standing efforts with Pt alloys, the desired improvement is yet to be achieved.

6.5.2 Core–Shell (CS) Structure

In electrocatalysis, stability of the catalyst is of paramount importance. Pt is the most stable and active electrocatalyst material for ORR. But on a cost basis, Pt is not favored and alternatives that combine activity, stability and cost criterion are highly desired. Core–shell structure of materials is reported for various applications and this concept can be highly favorable in ORR catalysis [39–41]. By the appropriate choice of the core and shell components, core–shell catalysts can improve the ORR activity, impart the required stability, and reduce the cost. Thus, monolayer coverage of ORR active material on suitable substrate surface is reported to enhance ORR activity because of the lattice strain and strain-induced electronic effects [42]. The shell should be active towards ORR and stable in fuel cell relevant operating conditions. The core can be anything that can electronically interact with the shell to cause the desired electronic effects; it is preferable to have a core that meets the cost criterion. The choice of core is also decided by their possible synthesis in the nanometer range. There are several reports on the synthesis and electrocatalytic activity of precious metal-based core–shell bimetallic catalysts [42–44]. The shell is typically Pt and the core preferably a TM element. Restricting the precious metal content to a monolayer coverage on the nanoparticle surface will significantly reduce its usage and increase the stability of the catalyst by preventing non-precious metal from leaching in harsh acidic electrochemical environment. Among the core–shell materials, the most investigated is the Pd-Pt model system (Fig. 6.6).

While it is relatively easy to deposit monolayer of the active material on suitable bulk substrate, it is a challenge to produce a monolayer-thick shell on nanomaterials of size 5–10 nm. Core–shell structures reported in the literature are produced mainly by three methods: (a) surface segregation of precious metal in bulk Pt_3M alloys at high temperature, (b) galvanic displacement of under-potential deposited copper (Cu_{upd}) monolayer by Pt on precious metals like Pd [30, 41, 43], and (c) electroless deposition of Cu and subsequent replacement of the deposited Cu with Pt. Each of these methods has its own limitations. While replacement of under potential deposited Cu by Pt on Pd is widely reported, often deposition of Cu occurs at various potentials on various low-index crystallographic orientations present on the polycrystalline Pd nanoparticles. In general, the thermodynamic reversible potential of Cu deposition is 0.34 V, but deposition on some of the selected orientations occurs even at very high potentials; for example, at 0.53 and 0.65 V, respectively, on Pd (100) and Pd (111) surfaces. Thus, multilayer deposition of Cu may occur on these surfaces at

0.34 V. Segregation of Pt by heat-treatment of the alloys (Pt-TM) and subsequent dealloying does not result in complete monolayer coverage of Pt on the TM core. Thus, an appropriate method still remains elusive for the controlled growth of monolayer shell on suitable core. The method should be scalable and should result in substantial reduction of the precious metal content without compromising the activity and stability of the catalyst (Table A1). Currently, the core–shell structures reported for electrocatalysis have a Pt:TM atomic ratio of 1:1 to 1:3, which is still a significant precious metal content. Core–shell catalysts with minimum shell thickness (monolayer) can greatly reduce the Pt content.

Sophisticated characterization techniques are available for the physical characterization of these materials. The techniques involve line scan, elemental mapping, selected area electron diffraction (SAED), and scanning transmission electron microscopy (STEM) [45, 46]. These methods often give bulk as well as individual particle composition. But it is often tedious to establish monolayer coverage of the shell on the core. Appropriate electrochemical methods for the confirmation of the core–shell structure are lacking in the literature. Overall, the concept of core–shell catalysts has generated sufficient enthusiasm among the electrochemical research community. But much needs to be done before such ideas turn to practical catalysts of importance. This will involve making a judicious choice of the core and shell, establishing the structure through appropriate characterization, and the scale-up of the synthesis methods [44]. The activity of CS catalysts is compared on a Pt metal content basis and on the basis of total precious metal content. Of course, the activity of CS catalysts is higher than that of the alloys, but their stability and long-term performance are yet to be established (Fig. 6.6).

6.5.3 Shape-Controlled Nanoparticles

Nanoparticles, with at least one of its dimensions in the range of 1–100 nm, are of fundamental importance in modern science and technology. The characteristics of such metal nanoparticles are decided by the shape, size, surface structure, and composition. Among these parameters, shape of the nanoparticle has a profound impact on the catalytic and electrocatalytic properties. Thus,

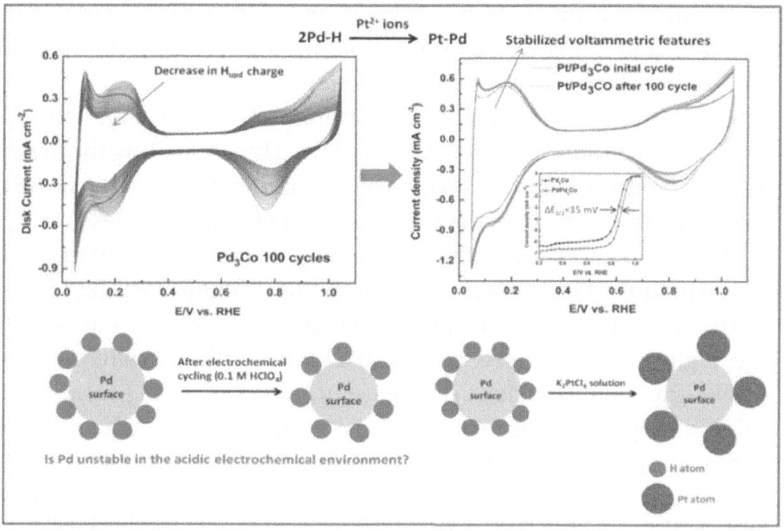

Figure 6.6 Pd_{core}-Pt_{shell} catalyst synthesized by exploiting the adsorbed hydrogen on the surface of Pd-based catalysts. Reproduced from Singh et al. [44] by permission of the PCCP Owner Societies. Copyright © 2013 Royal Chemical Society.

extensive research efforts have been devoted on the shape control of Pt, Pd, Ag, and Au nanoparticles [47–49]. The surface of these shape-controlled nanoparticles is usually terminated by the low index crystallographic orientations (Fig. 6.7).

From the single crystal studies, it is understood that the surface properties are dependent on the type of the exposed crystallographic orientations. Thus, in electrocatalysis, activities of the three dominant low-index facets—{111}, {100}, and {110}—are not the same (Table A1). For example, the ORR activity in sulfuric acid solution is in the order of (110) > (100) > (111). The orientation-dependent activity is due to the difference in O_2 adsorption energies of various crystallographic planes. In alkaline solution, this order is (111) > (110) > (100), and in perchloric acid solution it is (110) > (111) > (100) [3, 4, 50]. Thus, the order of activity also depends on the nature of the electrolyte and this is due to the interfering influence of the adsorbed anions from the electrolyte medium. According to Sabatier's principle, if the adsorption energy of O_2 is too high the catalyst surface will form an oxide, and if it is weak, oxygen will not interact with the catalyst surface. The highest activity is obtained when the adsorption energy of O_2 on the catalyst surface is optimum. These observations on bulk single crystal surfaces can be exploited by populating the highly active orientation on the surface of the nanoparticles. This in turn can be achieved through the shape-control of nanoparticles.

Metal nanoparticles minimize their surface free energy through the formation of low-index crystallographic orientations, viz., $\{100\}$, $\{110\}$ and $\{111\}$, with $\{111\}$ having the lowest and $\{110\}$ the highest surface free energy [51]. Generally, tetrahedral and octahedral nanoparticle surfaces are occupied by $\{111\}$ facets, and cubic nanoparticle surface with $\{100\}$ facets. Cuboctahedral has both $\{111\}$ and $\{100\}$ orientations on the surface. Hence, shape-controlled nanoparticles have received much research attention and they are synthesized using surfactants, capping agents, and stabilizers. In spite of their technical and fundamental importance, the exploitation of the shape-controlled nanoparticles is limited by the difficulty associated with their synthesis and cleaning [52].

Adsorptive capping agents are required in controlling the shape since they play a pivotal role in controlling the growth kinetics of nanoparticles in a particular direction and in hindering the growth in the other directions [51]. Since various facets have different energies, a given capping agent adsorbs selectively on the exposed facets. The preferential adsorption on a given set of surface orientations leads to anisotropic growth rates among the various crystallographic planes [51]. Thus, shape of the nanoparticle can be controlled using the capping agents.

Besides organic and inorganic molecules, many other chemicals including inorganic ions (e.g., halides) can also be used to control the natural growth habits of the nanoparticles [12, 53]. In principle, through judicious choice of the reagents and experimental conditions, any predetermined shape with selected low-index facets may be obtained. The nanoparticles with smaller size are prone to get in contact with other crystals leading to agglomeration of particles; consequently, the catalytic activity decreases. Therefore, one of the challenging issues during the synthesis of nanoparticles is the

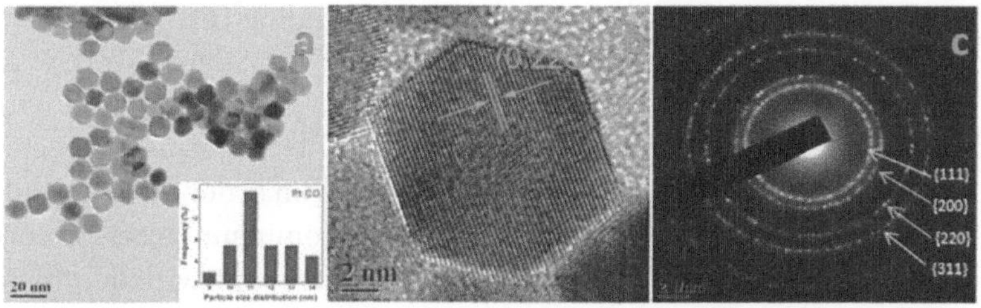

Figure 6.7 (a) TEM image of Pt-CO nanoparticles, (b) HRTEM image, and (c) and SAED pattern. Inset to (a) shows the particle size distribution histogram. Reprinted with permission from Devivaraprasad et al. [56]. Copyright © 2014 American Chemical Society.

monodispersity. The high energy surfaces of the metal nanoparticles are stabilized using hydrophobic organic stabilizers [51].

After the synthesis of the shape-controlled catalyst, the removal of stabilizers and capping agents is difficult as they adsorb strongly to the surface, blocking the active sites of nanoparticles. A proper cleaning procedure for the removal of stabilizers from the catalyst surface without compromising the size, shape and surface-structure has been a challenging issue for the research community [54, 55].

The synthesis of shape-controlled Pt nanoparticles such as Pt nanocubes with dominant (100) sites and Pt octahedral with dominant (111) sites is reported in the literature and their ORR activity is in line with that of the Pt single crystal surfaces, i.e., (100)-dominated Pt nanocube is more active than the (111)-dominated Pt octahedral in 0.5 M H_2SO_4 (Fig. 6.7). The order of ORR activity of the shape-controlled Pt nanoparticles depends on the electrolyte medium; thus, the ORR activity order in 0.1 M $HClO_4$ solution is $PtNP_{hexa}$ ((100) and (111))>$PtNP_{tetra}$ (111) >$PtNP_{sphe}$ (no preferential sites) >$PtNP_{cubic}$ (100), and in 0.5 M H_2SO_4 solution, it is $PtNP_{hexa}$ ((100) and (111))>$PtNP_{cubic}$ (100)>$PtNP_{sphe}$ >$PtNP_{tetra}$ (111). Overall, the $PtNP_{tetra}$ (111) has low activity in sulfuric acid but high activity in perchloric acid, whereas, $PtNP_{cubic}$ (100) has low activity in perchloric acid but high activity in sulfuric acid medium. $PtNP_{hexa}$ benefits from having both type of sites (Table A1) [56, 57].

In the literature, the shape-controlled Pt-alloy (Pt-Ni) nanoparticles are reported for the elctrocatalysts of ORR. Stamenkovic et al. reported a ~100 mV upshift in the half-wave potential of ORR with Pt_3Ni (111) when compared to that of Pt (111); in terms of specific activity at 0.9 V, the improvement amounts to an order of magnitude [34]. Pt-Ni of octahedral shape with (111) dominant facets is more active when compared to that of Pt-Ni of cubic shape with dominant (100) sites. A ~60 mV improvement in terms of half-wave potential was observed with the Pt_3Ni of (111)-dominated octahedral shape when compared to Pt_3Ni of (100)-dominated cubic shape [13]. Methanol and ethanol oxidation activities are also found to be dependent on crystal orientation; the alcohol oxidation activity and onset potential are higher on Pt nanocube when compared to that of polycrystalline Pt [58]. The shape-controlled Pd metal particles are extensively used to verify the orientation-dependent oxidation of small organic molecules in acidic medium (0.1 M $HClO_4$). Thus, Jin et al. reported clear evidence for the facet-dependant activity towards formic acid oxidation; maximum current density is obtained with the cubic nanoparticles and minimum with octahedral nanoparticles [59].

The catalytic behavior also depends on the surface defect-structure such as high-index planes of the material that has good activity as compared to the low-index planes; in the former case the density of atomic steps, kinks, and edges, which usually act as more active sites for disrupting the chemical bonds, are more than that on the latter. For example, fundamental studies of Pt bulk single crystal materials with high-index planes exhibit higher catalytic activity when compared to that of the low-index {111}, {110}, and {100} planes [60]. Thus, the Pt {210} surface has higher CO_2 reduction capability as compared to that of the crystals with low-index planes, and Pt {410} facets have higher catalytic activity towards decomposition of NO compared to that with polycrystalline Pt and Pt {100} surfaces [61, 62]. The difficulty with preparation of nanoparticles with high-index planes is that they have higher surface energy when compared to that of the low-index planes. Effect of surface defect structure was reported for various fuel cell relevant reactions [63, 64]. In practice, they are reported with dendrite structures [65].

6.5.4 Pd-Based Alloy Catalysts

In an attempt to replace expensive Pt, several other precious metals were tested as ORR catalysts. Among them, Pd is the most promising candidate though the half-wave potential of ORR is ~100 mV less than that of Pt; moreover, Pd is less expensive and more abundant. Pd-based alloys are reported to have significant ORR activity, almost comparable to that of Pt (Table A1) [66]. The original work on the activity improvement with Pd-based systems was reported with sputter-deposited Pd alloy thin-film catalysts by Savadogo et al. [66]. Since then, these alloys have been prepared by a variety of synthesis procedures [29, 66–71]. Some of the methods include sol-gel, colloidal, sonochemical, solution-phase simultaneous reduction, and heat treatment of Pd/C with transition metal salts [68, 71–75]. A heat treatment of the catalyst in reducing atmosphere is followed in the preparation of the catalysts [29, 72, 74]. Among Pd-based catalysts, alloys with TM (Cr, Co, or Fe) have attracted significant research interest [29, 66, 76]. Moreover, Pd-based catalysts are tolerant towards methanol, which is of significant importance in the direct methanol fuel cells (DMFCs) [68, 69, 72]. The origin of activity improvement and stability concerns with Pd-based catalyst have been a subject of immense research activity [29, 44, 77, 78].

In the literature, the ORR activity improvement on PdM alloys (where M is a transition metal like Fe, Co, Ni, etc.) as compared to that of pure Pd is mainly explained on the basis of the formation of a more favorable Pd-Pd bond distance for the dissociative-adsorption of oxygen and its subsequent reduction to water [29]. Shao et al. synthesized PdFe alloy-based catalyst of an atomic composition of 3:1 (Pd:Fe) with highest activity and lowest Pd-Pd bond distance [29]. Several groups have proposed the theory of lattice contraction and consequent changes in electronic properties related to the compressive strain and ligand effects as a cause of improved activity of alloy catalysts [42, 69, 74]. However, Neergat et al. synthesized PdFe catalysts without involving any heat treatment procedure, and showed activity comparable to that of the heat-treated catalysts reported in the earlier studies, but with lattice expansion relative to Pd [68]. The role of OH formation on Pd has attracted significant research interest and it is reported to have an inhibiting effect on ORR. Recently, Rahul et al. showed activity improvement in case of electrochemically reduced Pd oxide-based catalysts [79]. The increase of the activity upon oxidative heat-treatment may be attributed to the increase in electrochemical surface area (ESA), change in the degree of alloying, segregation of the alloy components, changes in the particle composition, formation of a chemically ordered alloy phase, and increase in the size of Pd nanoparticles [79]. Rahul et al. further pointed out the active role of adsorbed-oxygenated species in the ORR [79]. An alternative explanation for the ORR activity improvement on Pd alloys is based on Gibbs free energies of the key elementary steps in electroreduction of oxygen (thermodynamic guidelines) [76]. On the basis of the thermodynamic principle, Fernandez et al. explained the role of bimetallic surfaces in the ORR mechanism [76]. This mechanism involves combining one metal (M) which is good in breaking the O–O bond with another metal (M′) which can catalyze the reduction of adsorbed oxygen atom. In case of Pd-Co alloy catalysts, Co facilitates the adsorption and dissociation of molecular oxygen and Pd effectively catalyzes the reduction of adsorbed oxygen [76].

It is found that the oxidative dissolution of Pd and leaching of the TM are the major causes of instability of Pd-based catalysts [44, 78]. There are several reports on improving the stability of Pd-based materials by coating with Pt; these core–shell structures help to improve stability by protecting the Pd core [42–44, 78]. Recently, Singh et al. demonstrated improved stability of Pd and Pd_3Co coated with half a monolayer of Pt during potential cycling in 0.1 M $HClO_4$ (Fig. 6.6) [44]. The Pt-coated Pd and Pd_3Co were synthesized by a simple deposition of Pt with adsorbed hydrogen on the Pd surface [44].

6.5.5 Non-precious Metal Electrocatalysts (NPMCs)

Another economically viable approach is to replace the highly expensive precious metal catalysts with cheap and abundant transition metal-based catalysts. The activity loss due to this transition can be eliminated by loading reasonably higher amount of the catalyst, instead of limiting the loading to a very low level as in the case of Pt. The idea of non-precious metal catalysts for ORR is adopted from the nature, where the reduction of oxygen is very effectively catalyzed by the enzyme cytochrome C oxidase [80]. The active center of cytochrome C oxidase is a multinuclear redox center. The initial approach toward NPMCs was based on similarly coordinated molecules that can mimic the biological system. Thus, since the report on ORR electrocatalysis with cobalt phthalocyanine in 1964 by Jasinski et al. [15], several alternative synthetic methods have been investigated to produce practical NPMCs [16–19, 81, 82]. In general, fuel cells are operated in alkaline and acidic conditions. Thus, it is important to design the electrocatalyst that would catalyze the ORR efficiently in the desired pH range. Most of the results reported in the literature with NPMCs for ORR have been in alkaline media. This is mainly because of the poor stability of the transition metal-based catalysts in acidic media. Moreover, kinetics of ORR is more facile with transition metal (TM) in alkaline medium than that in acidic medium. Therefore, NPMCs can be highly effective in catalyzing ORR in the former medium. Thus, the development of alkaline fuel cells (AFCs) tremendously encourages the use of NPMCs. Though the stability criterion alone negates any possible use of NPMCs in acidic media, limited success (in terms of both stability and activity) has been reported with TM-metal-based catalysts, i.e., nitrogen coordinated TM (Fe, Co, and Ni) bound to an aromatic carbon structure [19]. In the initial work, the catalysts were obtained by pyrolyzing Fe-N_4 or Co-N_4 macrocycles adsorbed on a carbon-support in an inert atmosphere [18, 19, 81, 82]. The TMs generally used in NPMCs are Fe, Co, Ni, Mn, etc., and carbon originates from the pyrolysis of nitrogen-containing multi-dentate macrocyclic organic ligands attached to the central metal atom [81, 83, 84]. Pyrolyzed TM-based porphyrines and pthalocyanines constitute examples of such NPMCs [16, 81, 83–85]. Superior activity has been reported with these materials subjected to high-temperature annealing that is sufficient to remove hydrogen and most of the nitrogen (relative to their non-heat-treated counterparts). But the origin of the catalytic activity is debatable [86]. Gupta et al. substituted these expensive pyrolyzed macrocycle-based catalysts with those derived from individual TM and nitrogen precursors [87]. Dodelet and coworkers reported materials with improved activity and stability (24 h in a

PEMFC) synthesized by the pyrolysis of carbon, ammonia, and iron precursor (iron acetate [FeAc]) [19]. The two pore fillers used were nitrogen-free perylene-tetracarboxylic dianhydride (PTCDA), and nitrogen-containing 1, 10-phenanthroline (phen). With nitrogen-free PTCDA, the catalyst was pyrolyzed in NH_3 atmosphere to incorporate the N sites in the carbon, while with nitrogen-containing phen, the catalyst was pyrolyzed in either Ar or NH_3 atmosphere [19]. The activity of these NPMCs is reported to have improved by a factor of 35 relative to that of the previous generation of Fe-based NPMCs. But, due to thicker catalyst layer (5.3 mg cm^{-2} loading of NPMC when compared to 0.4 mg cm^{-2} of Pt), high mass transport losses were incurred at high current density [19]. Using a variety of characterization techniques, including secondary ion mass spectroscopy, Lefevre et al. established that Fe-coordinated to two pyridinic nitrogen groups at the edge of graphene sheets as the active sites [88]. But the remaining coordination of the central metal atom remains unknown. The exact mechanism and active center is not clear, but graphene-coordinated FeN_4 or FeN_2 moieties were proposed [18, 19, 89, 90]. The replacement of precious metal with NPMCs reduces the cost of the fuel cell stack to a great extent alongside catalyzing the fuel cell reactions almost as efficiently as Pt. Bashyam et al. demonstrated a Co-polypyrrole composite (Co-PPy-C) by the chemical treatment of Co-polypyrrole with carbon in inert atmosphere [18]. This procedure is devoid of the pyrolysis step unlike those used for the synthesis of other NPMCs so far. The catalyst is reported to exhibit good activity and stability, with a Co loading of 0.06 mg cm^{-2} in both H_2–O_2 and H_2–air PEFCs [18]. The proposed reason for this improvement is the entrapment of Co sites into the polypyrrole matrix and formation of CoN_x active sites resulting from the strong Co–polypyrrole interactions. Macrocyclic N_4 complexes with polyaniline, polyporphirine, and tetraazzulenes were synthesized and the Me–N_4 complexes were used as the catalyst in the past [18, 84, 91]. Among NPMCs, Fe–N_4 and Co–N_4 complexes, where the central transition metal ion is coordinated with nitrogen atoms, have shown highest electrocatalytic activity towards ORR [19, 92].

Studies reveal that heat treatment of carbon-based materials with different dopants like nitrogen, sulfur, and boron effectively enhances ORR activity of such transition metal macrocyclic complexes

[93–96]. Heat treatment at different temperatures forms pyridinic, pyrolic and quartenary nitrogen coordinated with the transition metal center, which acts as an active site for ORR. With the advent of graphene oxide and graphene recently, their exotic properties have been investigated for improved ORR performance in alkaline medium. These sp^2 hybridized allotropes of carbon doped with nitrogen (using ammonia precursor), polyaniline (PANI), and polypyrrole (PPy) are reported to exhibit improved ORR activity [97].

The graphene–carbon nitride (G-CN) composite has been recently reported to exhibit improved ORR activity [98]. The higher nitrogen content in the graphene structure affects the limiting current density of ORR (along with ORR onset potential) and facilitates the $4e^-$ pathway during ORR instead of $2e^-$ pathway. Graphene doped with nonmetals (boron, sulfur, iodine, boron-sulfur) and with non-precious metals like Co and Fe is reported to have improved ORR activity in alkaline medium [93–96].

One of the major issues that deserve further attention with NPMCs is their poor turn over frequency (ToF, number of electrons transferred per active site per second), typically, 0.4 s^{-1}. The ToF of NPMCs has improved over time, and with recent candidate materials, values close to that of Pt/C (25 s^{-1}) have been reported [11, 19, 89]. But, the active site density of NPMCs has been found to be very poor. This disadvantage can partly be overcome by increasing the catalyst loading, but it results in serious mass transfer losses at the electrode. Therefore, the activity comparison is often made on a mA cm^{-3} basis instead of mA cm^{-2} [19]. The active site in NPMC is the transition metal site that is coordinated with four aromatic nitrogen functionalities [19]. Such a configuration of $TM/N_4/C_x$ will certainly reduce the active site density and further efforts are required to achieve a decent enough ToF and active site density with NPMCs.

6.6 Conclusions

Recent developments in nanomaterials for electrocatalysis of ORR in the last few years have been reviewed. Primarily, these electrocatalysts consist of supported alloys, core–shell materials, shape-controlled materials, and non-precious metal catalysts. Overall, the main objectives are to reduce the cost and to improve the activity and durability of the catalysts. Significant advances have been made toward achieving these goals with both precious metal and non-precious metal catalysts. Among the precious metal catalysts, the core–shell structure seems to be a promising option. With non-precious metal catalysts, there is a need to improve their active site density, ToF and most importantly the stability. The widespread use of RDE/RRDE, instead of fuel cell configuration, for the electrochemical characterization of the catalysts suggests a paradigm shift in research direction. Indirectly, this would mean that the current limits of activity, stability, and cost of the component materials needs to be further pushed ahead before an affordable fuel cell is available in the market.

Acknowledgments

The authors would like to acknowledge the Ph.D. students Rahul Ramesh, Naresh Nalajala, Ruttala Devivaraprasad, Tathagata Kar, Bapi Bera, and Arup Chakraborty for their useful comments and discussion.

Appendix

Table A1 Specific and mass activities of Pt-based catalysts

Catalyst	Specific activity (mA/cm^2)		Mass activity (mA/mg)		Experimental conditions
	0.85 V	0.9 V	0.85 V	0.9 V	
Platinum single crystal and Pt unsupported catalysts (black)					
Pt {100}	1.450	0.480			In 0.05 M H_2SO_4 at 60°C [99]
Pt {110}	5.400	1.900			
Pt {111}	0.650	0.080			
Pt polycrystalline (pc)	0.4	0.110			In 0.5 M H_2SO_4 at 60°C [1]

(*Contd.*)

Table A1 (*Contd.*)

Pt (pc)	3	0.9			In 0.1 M $HClO_4$ at 60°C [31]
Pt-poly		1.4			In 0.1 M $HClO_4$ at 60°C [30]
5 mm Ptpc disk		2.770			In 0.1 M $HClO_4$ at 60°C [11]
Pt black (HiSpec 1000, JM)		0.860		50	
Pt black		0.25		50	In 0.1 M $HClO_4$ at room temperature [12]
Pt black		0.2		50	In 0.1 M $HClO_4$ at room temperature [14]
Pt polycrystalline	1.02	0.3	246	68.68	In 0.1 M $HClO_4$ at room temperature [56]
Pt nanocube	0.95	0.32	190.7	64.6	
Pt tetrahedral	1.5	0.4	348	88	
Pt cuboctahedral	0.56	0.31	114	67	
Pt-based alloys (single crystal and unsupported)					
Pt_3Ni (Skin)		3.3			In 0.1 M $HClO_4$ at 60°C [30]
Pt_3Fe (Skin)		3.9			
Pt_3Co (Skin)		4.3			
Pt_3Ni (pc)	6.25	1.25			In 0.1 M $HClO_4$ at 60°C [31]
Pt_3Co (pc)	8	1.4			
PdPt Dendrites		0.42		200	In 0.1 M $HClO_4$ at room temperature [12]
PtCu alloy		0.495		150	In 0.1 M $HClO_4$ at room temperature [14]
Cu-Pt core–shell		0.7		260	
Supported platinum and platinum based alloy					
40% Pt/Vulcan (ETEK)		0.320		98	In 0.1 M $HClO_4$ at 60°C [11]
20% Pt/Vulcan (ETEK)		0.300		200	
45.9% Pt/HSC-E (TKK)		0.290		220	
46.5% Pt/HSC-E (TKK)		0.250		200	
27% Pt/Ketjen		0.160		200	

(*Contd.*)

Table A1 (*Contd.*)

Catalyst	Specific activity (mA/cm^2)		Mass activity (mA/mg)		Experimental conditions
	0.85 V	0.9 V	0.85 V	0.9 V	
Pt/Vulcan H_2SO_4	0.19	0.035	–	–	In 0.5 M H_2SO_4 at 60°C [1]
Pt/Vulcan $HClO_4$	0.51	0.065	–	–	
Pt/C	–	–	200	–	In 0.1 M $HClO_4$ at room temperature [41]
Pt (6 nm)/C	–	0.97	–	374	In 0.1 M $HClO_4$ at 60°C [26]
Pt/C	0.062	0.019	–	–	In 1 M $HClO_4$ at room temperature [100]
$Pt_{ml}/PdCo_5$	–	–	800	–	In 0.1 M $HClO_4$ at room temperature [41]
$Pt_{ml}/PtCo_5$	–	–	1200	–	
Pt_3Co (9 nm)	–	3.43	–	998	In 0.1 M $HClO_4$ at 60°C [26]
Pt_3Co (6 nm)	–	3.14	–	1233	
Pt_3Co (3 nm)	–	1.62	–	1176	
PtCo	0.319	0.116	–	–	In 1 M $HClO_4$ at room temperature [100]
Pt/Vulcan	–	–	900	200	In 0.1 M $HClO_4$ at 60°C [31]
Pt_3Ni/Vulcan	–	–	1200	250	
Pt_3Co/Vulcan	–	–	1250	250	
PtCo/C	–	0.625	–	–	
Palladium and palladium based alloys (unsupported and supported)					
Pd/C	0.131				In 0.1 M $HClO_4$ at room temperature [29]
Pd_3Fe/C	0.791				
Pd/C As-prepared	0.19	0.03	81.3	18	In 0.1 M $HClO_4$ at room temperature [79]
Pd/C OHT	0.17	0.03	94	19.1	
Pd_3Co/C As-prepared	0.41	0.09	367.4	90.4	
Pd_3Co/C OHT	0.45	0.12	452.2	130.6	

References

1. Paulus, U. A., Schmidt, T. J., Gasteiger, H. A., and Behm, R. J. (2001) Oxygen reduction on a high-surface area Pt/Vulcan carbon catalyst: a thin-film rotating ring-disk electrode study, *J. Electroanal. Chem.*, **495**, 134–145.
2. Wilson, M. S., Garzon, F. H., Sickafus, K. E., and Gottesfeld, S. (1993). Surface area loss of supported platinum in polymer electrolyte fuel cells, *J. Electrochem. Soc.*, **140**, 2872–2877.
3. Markovic, N. M., Gasteiger, H. A., and Jr. Ross, P. N. (1995). Oxygen reduction on platinum low-index single-crystal surfaces in sulfuric acid solution: rotating ring-Pt(hkl) disk studies, *J. Phys. Chem.*, **99**, 3411–3415.
4. Markovic, N. M., Gasteiger, H. A., and Jr. Ross, P. N. (1996). Oxygen reduction on platinum low-index single-crystal surfaces in alkaline solution: rotating ring disk Pt(hkl) studies, *J. Phys. Chem.*, **100**, 6715–6721.
5. Dinh, H. N., Ren, X., Garzon, F. H., Zelenay, P., and Gottesfeld, S. (2000). Electrocatalysis in direct methanol fuel cells: in-situ probing of PtRu anode catalyst, *J. Electroanal. Chem.*, **491**, 222–233.
6. Green, C. L. and Kucernak, A. (2002). Determination of the platinum and ruthenium surface areas in platinum-ruthenium alloy electrocatalysts by underpotential deposition of copper. I. unsupported catalysts, *J. Phy. Chem. B*, **106**, 1036–1047.
7. Wroblowa, H. S., Yen, C. P., and Razumney, G. (1976). Electroreduction of oxygen: a new mechanistic criterion, *J. Electroanal. Chem.*, **69**, 195–201.
8. Springer, T. E. and Raistrick, I. D. (1989). Electrical impedance of a pore wall for the flooded-agglomerate model of porous gas-diffusion electrodes, *J. Electrochem. Soc.*, **136**, 1594–1603.
9. Springer, T. E., Zawodzinski, T. A., Wilson, M. S., and Gottesfeld, S. (1996). Characterization of polymer electrolyte fuel cells using AC impedance spectroscopy. *J. Electrochem. Soc.*, **143**, 587–599.
10. Perez, J., Gonzalez, E. R., and Ticianelli, E. A. (1998). Impedance studies of the oxygen reduction on thin porous coating rotating platinum electrodes, *J. Electrochem. Soc.*, **145**, 2307–2313.
11. Gasteiger, H. A., Kocha, S. S., Sompalli, B., and Wagner, F. T. (2005). Activity benchmarks and requirements for Pt, Pt-alloy, and non-Pt oxygen reduction catalysts for PEMFCs, *App. Catal. B: Environ.*, **56**, 9–35.

12. Lim, B., Jiang, M., Camargo, P. H. C., Cho, E. C., Tao, J., Lu, X., Zhu, Y., and Xia, Y. (2009). Pd-Pt bimetallic nanodendrites with high activity for oxygen reduction, *Science*, **324**, 1302–1305.
13. Zhang, J., Yang, H., Fang, J., and Zou, S. (2010). Synthesis and oxygen reduction activity of shape-controlled Pt_3Ni nanopolyhedra, *Nano Lett.*, **10**, 638–644.
14. Neergat, M. and Rahul, R. (2012). Unsupported Cu-Pt core–shell nanoparticles: oxygen reduction reaction (ORR) catalyst with better activity and reduced precious metal content, *J. Electrochem. Soc.*, **159**, F234–F241.
15. Jasinski, R. (1964). A new fuel cell cathode catalyst, *Nature*, **201**, 1212–1213.
16. Sun, G. Q., Wang, J. T., and Savinell, R. F. (1998). Iron(III) tetramethoxyphenylporphyrin (FeTMPP) as methanol tolerant electrocatalyst for oxygen reduction in direct methanol fuel cells, *J. Appl. Electrochem.*, **28**, pp.1087–1093.
17. Feng, Y. and Alonso-Vante, N. (2008). Nonprecious metal catalysts for the molecular oxygen-reduction reaction, *Phys. Status Solidi B*, **245**, 1792–1806.
18. Bashyam, R. and Zelenay, P. (2006). A class of non-precious metal composite catalysts for fuel cells, *Nature*, **443**, 63–66.
19. Lefèvre, M., Proietti, E., Jaouen, F., and Dodelet, J.-P. (2009). Iron-based catalyst with improved oxygen reduction activity in polymer electrolyte fuel cells, *Science*, **324**, 71–74.
20. Yu, D., Zhang, Q., and Dai, L. (2010). Highly efficient metal-free growth of nitrogen-doped single-walled carbon nanotubes on plasma-etched substrates for oxygen reduction, *J. Am. Chem. Soc.*, **132**, 15127–15129.
21. Li, Y., Zhou, W., Wang, H., Xie, L., Liang, Y., Wei, F., Idrobo, J. C., Pennycook, S. J., and Dai, H. (2012). An oxygen reduction electrocatalyst based on carbon nanotube–graphene complexes, *Nat. Nanotech.*, **7**, 394–400.
22. Salgado, J. R. C., Antolini, E., and Gonzalez, E. R. (2005). Carbon supported Pt–Co alloys as methanol-resistant oxygen-reduction electrocatalysts for direct methanol fuel cells. *Appl. Catal. B*, **57**, 283–290.
23. Neergat, M., Seiler, T., Savinova, E. R., and Stimming, U. (2006). Improvement of the performance of a direct methanol fuel cell using a pulse technique, *J. Electrochem. Soc.*, **153**, A997–A1003.

24. Debe, M. K. (2012). Electrocatalyst approaches and challenges for automotive fuel cells, *Nature*, **486**, 43–51.

25. Nesselberger, M., Ashton, S., Meier, J. C., Katsounaros, I., Mayrhofer, K. J. J., and Arenz, M. (2011). The particle size effect on the oxygen reduction reaction activity of Pt catalysts: influence of electrolyte and relation to single crystal models, *J. Am. Chem. Soc.*, **133**, 17428–17433.

26. Wang, C., Vliet, D. v. d., Chang, K.-C., You, H., Strmcnik, D., Schlueter, J. A., Markovic, N. M., and Stamenkovic, V. R. (2009). Monodisperse Pt_3Co nanoparticles as a catalyst for the oxygen reduction reaction: size-dependent activity, *J. Phys. Chem. C*, **113**, 19365–19368.

27. Alonso, F. J. P., McCarthy, D. N, Nierhoff, A., Fernandez, P. H., Strebel, C., Stephens, I. E. L., Nielsen, J. H., and Chorkendorff, I. (2012). The Effect of size on the oxygen electroreduction activity of mass selected platinum nanoparticles, *Angew. Chem. Int. Ed.*, **51**, 4641–4643.

28. Solla-Gullón, J., Rodriguez, P., Herrero, E., Aldaz, A., and Feliu, J. M. (2008). Surface characterization of platinum electrodes, *Phys. Chem. Chem. Phys.*, **10**, 1359–1373.

29. Shao, M.-H., Sasaki, K., and Adzic, R. R. (2006). Pd-Fe Nanoparticles as electrocatalysts for oxygen reduction, *J. Am. Chem. Soc.*, **128**, 3526–3527.

30. Stamenkovic, V. R., Mun, B. S., Mayrhofer, K. J. J., Ross, P. N., and Markovic, N. M. (2006). Effect of surface composition on electronic structure, stability, and electrocatalytic properties of Pt-transition metal alloys: Pt-skin versus Pt-skeleton surfaces, *J. Am. Chem. Soc.*, **128**, 8813–8819.

31. Paulus, U. A., Wokaun, A., Scherer, G. G., Schmidt, T. J., Stamenkovic, V., Markovic, N. M., and Ross, P. N. (2002). Oxygen reduction on high surface area Pt based alloy catalysts in comparison to well defined smooth bulk alloy electrodes, *Electrochim. Acta*, **47**, 3787–3798.

32. Mukerjee, S., Srinivasan, S., Soriaga, M. P., and McBreen, J. (1995). Role of structural and electronic properties of Pt and Pt alloys on electrocatalysis of oxygen reduction: an in situ XANES and EXAFS investigation, *J. Electrochem. Soc.*, **142**, 1409–1422.

33. Shukla, A. K., Neergat, M., Bera, P., Jayaram, V., and Hegde, M. S. (2001). An XPS study on binary and ternary alloys of transition metals with platinized carbon and its bearing upon oxygen electroreduction in direct methanol fuel cells, *J. Electroanal. Chem.*, **504**, 111–119.

34. Stamenkovic, V. R., Fowler, B., Mun, B. S., Wang, G., Ross, P. N., Lucas, C. A., and Markovic, N. M. (2007). Improved oxygen reduction activity on $Pt_3Ni(111)$ via increased surface site availability, *Science*, **315**, 493–497.
35. Neergat, M. and Shukla, A. K. (2001). A high-performance phosphoric acid fuel cell, *J. Power Sources*, **102**, 317–321.
36. Jalan, V. and Taylor, E. J. (1983). Importance os interatomic spacing in catalytic reduction of oxygen in phosphoric acid, *J. Electrochem. Soc.*, **130**, 2299–2302.
37. Paffett, M. T., Beery, J. G., and Gottesfeld, S. (1988). Oxygen reduction at $Pt_{0.65}Cr_{0.35}$, $Pt_{0.2}Cr_{0.8}$ and roughened platinum, *J. Electrochem. Soc.*, **135**, 1431–1436.
38. Stamenkovic, V., Schmidt, T. J., Ross, P. N., and Markovic, N. M. (2002). Surface composition effects in electrocatalysis: kinetics of oxygen reduction on well-defined Pt_3Ni and Pt_3Co alloy surfaces, *J. Phys. Chem. B*, **106**, 11970–11979.
39. Zhang, L., Blom, D. A., Wang, Z., and Wang, H. (2011). Au–Cu_2O core–shell nanoparticles: a hybrid metal–semiconductor heteronanostructure with geometrically tunable optical properties, *Chem. Mater.*, **23**, 4587–4598.
40. Chen, Y.-J., Gao, P., Wang, R. X., Zhu, C. L., Wang, L. J., Cao, M. S., and Jin, H. B. (2009). Porous Fe_3O_4/SnO_2 core/shell nanorods: synthesis and electromagnetic properties, *J. Phys. Chem. C*, **113**, 10061–10064.
41. Zhang, J., Lima, F. H. B., Shao, M. H., Sasaki, K., Wang, J. X., Hanson, J., and Adzic, R. R. (2005). Platinum monolayer on nonnoble metal-noble metal core–shell nanoparticle electrocatalysts for O_2 reduction, *J. Phys. Chem. B*, **109**, 22701–22704.
42. Zhang, J., Vukmirovic, M. B., Xu, Y., Mavrikakis, M., and Adzic, R. R. (2005). Controlling the catalytic activity of platinum monolayer electrocatalysts for oxygen reduction with different substrates, *Angew. Chem. Int. Ed.*, **44**, 2132–2135.
43. Taufany, F., Pan, C.-J., Rick, J., Chou, H.-L., Tsai, M.-C., Hwang, B.-J., Liu, D.-G., Lee, J.-F., Tang, M.-T., Lee, Y.-C., and Chen, C.-I. (2011). Kinetically controlled autocatalytic chemical process for bulk production of bimetallic core–shell structured nanoparticles, *ACS Nano*, **5**, 9370–9381.
44. Singh, R. K., Rahul, R., and Neergat, M. (2013). Stability issues in Pd-based catalysts: the role of surface Pt in improving the stability and

oxygen reduction reaction (ORR) activity', *Phys. Chem. Chem. Phys.*, **15**, 13044–13051.

45. Zhou, W.-P., Sasaki, K., Su, D., Zhu, Y., Wang, J. X., and Adzic, R. R. (2010). Gram-scale-synthesized Pd_2Co-supported Pt monolayer electrocatalysts for oxygen reduction reaction, *J. Phys. Chem. C*, **114**, 8950–8957.
46. Wang, J. X., Inada, H., Wu, L., Zhu, Y., Choi, Y. M., Liu, P., Zhou, W.-P., and Adzic, R. R. (2009). Oxygen reduction on well-defined core–shell nanocatalysts: particle size, facet, and Pt shell thickness effects, *J. Am. Chem. Soc.*, **131**, 17298–17302.
47. Xia, Y., Xiong, Y., Lim, B., and Skrabalak, S. E. (2009). Shape-controlled synthesis of metal nanocrystals: simple chemistry meets complex physics?, *Angew. Chem. Int. Ed.*, **48**, 60–103.
48. Lim, B., Jiang, M., Tao, J., Camargo, P. H. C., Zhu, Y., and Xia, Y. (2009). Shape-controlled synthesis of Pd nanocrystals in aqueous solutions, *Adv. Funct. Mater.*, **19**, 189–200.
49. Cheong, S., Watt, J. D., and Tilley, R. D. (2010). Shape control of platinum and palladium nanoparticles for catalysis, *Nanoscale*, **2**, 2045–2053.
50. Markovic, N. M., Adzic, R. R., Cahan, B. D., and Yeager, E. B. (1994). Structural effects in electrocatalysis: oxygen reduction on platinum low index single-crystal surfaces in perchloric acid solutions, *J. Electroanal. Chem.*, **377**, 249–259.
51. Peng, Z. and Yang, H. (2009). Designer platinum nanoparticles: control of shape, composition in alloy, nanostructure and electrocatalytic property, *Nano Today*, **4**, 143–164.
52. Tao, A. R., Habas, S., and Yang, P. (2008). Shape control of colloidal metal nanocrystals, *Small*, **4**, 310–325.
53. Xiong, Y., Cai, H., Wiley, B. J., Wang, J., Kim, M. J., and Xia, Y. (2007). Synthesis and mechanistic study of palladium nanobars and nanorods, *J. Am. Chem. Soc.*, **129**, 3665–3675.
54. Naresh, N., Wasim, F. G. S., Ladewig, B. P., and Neergat, M. (2013). Removal of surfactant and capping agent from Pd nanocubes (Pd-NCs) using tert-butylamine: its effect on electrochemical characteristics, *J. Mater. Chem. A*, **1**, 8553–8559.
55. Naresh, N., Wasim, F. G. S., Ladewig, B. P., and Neergat, M. (2014). Sodium borohydride treatment: a simple and effective process for the removal of stabilizer and capping agents from shape-controlled palladium nanoparticles, *Chem. Commun.*, **50**, 9365–9368.

56. Devivaraprasad, R., Rahul, R., Naresh, N., Kar, T., Singh, R. K., and Neergat, M. (2014). Oxygen reduction reaction and peroxide generation on shape-controlled and polycrystalline platinum nanoparticles in acidic and alkaline electrolytes, *Langmuir*, **30**, 8995–9006.

57. Sanchez-Sanchez, C. M. S., Solla-Gullon, J., Vidal-Iglesias, F. J., Aldaz, A., Montiel, V., and Herrero, E. (2010). Imaging structure sensitive catalysis on different shape-controlled platinum nanoparticles, *J. Am. Chem. Soc.*, **132**, 5622–5624.

58. Han, S. B., Song, Y. J., Lee, J. M., Kim, J. Y., and Park, K. W. (2008). Platinum nanocube catalysts for methanol and ethanol electrooxidation, *Electrochem. Commun.*, **10**, 1044–1047.

59. Jin, M., Zhang, H., Xie, Z., and Xia, Y. (2012). Palladium nanocrystals enclosed by {100} and {111} facets in controlled proportions and their catalytic activities for formic acid oxidation, *Energy Environ. Sci.*, **5**, 6352–6357.

60. Tian, N., Zhou, Z. Y., Sun, S. G., Ding, Y., and Wang, Z. L. (2007). Synthesis of tetrahexahedral platinum nanocrystals with high-index facets and high electro oxidation activity, *Science*, **316**, 732–735.

61. Hoshi, N., Kawatani, S., Kudo, M., and Hori, Y. (1999). Significant enhancement of the electrochemical reduction of CO_2 at the kink sites on Pt(S)-[*n*(110)x(100)] and Pt(S)-[*n*(100)x(110)], *J. Electroanal. Chem.*, **467**, 67–73.

62. Banholzer, W. F., and Masel, R. I. (1984). Nitric oxide decomposition on Pt(410), *J. Catal.*, **85**, 127–134.

63. Macia, M. D., Campina, J. M., Herrero, E., and Feliu, J. M. (2004). On the kinetics of oxygen reduction on platinum stepped surfaces in acidic media. *J. Electroanal. Chem.*, **564**, 141–150.

64. Spendelow, J. S., Xu, Q., Goodpaster, J. D., Kenis, P. J. A., and Wieckowski, A. (2007). The role of surface defects in CO oxidation, methanol oxidation, and oxygen reduction on Pt (111), *J. Electrochem. Soc.*, **154**, F238–F242.

65. Lim, B., Jiang, M., Camargo, P. H. C., Cho, E. C., Tao, J., Lu, X., Zhu, Y., and Xia, Y. (2009). Pd-Pt Bimetallic nanodendrites with high activity for oxygen reduction, *Science*, **324**, 1302–1305.

66. Savadogo, O., Lee, K., Oishi, K., Mitsushima, S., Kamiya, N., and Ota, K.-I. (2004). New palladium alloys catalyst for the oxygen reduction reaction in an acid medium, *Electrochem. Commun.*, **6**, 105–109.

67. Antolini, E. (2009). Palladium in fuel cell catalysis, *Energy Environ. Sci.*, **2**, 915–931.

68. Neergat, M., Gunasekar, V., and Rahul, R. (2011). Carbon-supported Pd–Fe electrocatalysts for oxygen reduction reaction (ORR) and their methanol tolerance, *J. Electroanal. Chem.*, **658**, 25–32.

69. Wang, W., Zheng, D., Du, C., Zou, Z., Zhang, X., Xia, B., Yang, H., and Akins, D. L. (2007). Carbon-supported Pd-Co bimetallic nanoparticles as electrocatalysts for the oxygen reduction reaction, *J. Power Sources*, **167**, 243–249.

70. Sarapuu, A., Kasikov, A., Wong, N., Lucas, C. A., Sedghi, G., Nichols, R. J., and Tammeveski, K. (2010). Electroreduction of oxygen on gold-supported nanostructured palladium films in acid solutions, *Electrochim. Acta*, **55**, 6768–6774.

71. Alexeyeva, N., Sarapuu, A., Tammeveski, K., Vidal-Iglesias, F. J., Solla-Gullón, J., and Feliu, J. M. (2011). Electroreduction of oxygen on Vulcan carbon supported Pd nanoparticles and Pd-M nanoalloys in acid and alkaline solutions, *Electrochim. Acta*, **56**, 6702–6708.

72. Zhang, L., Lee, K., Zhang, J. (2007). Effect of synthetic reducing agents on morphology and ORR activity of carbon-supported nano-Pd–Co alloy electrocatalysts, *Electrochim. Acta*, **52**, 3088–3094.

73. Godínez-García, A., Pérez-Robles, J. F., Martínez-Tejada, H. V., Solorza-Feria, O. (2012). Characterization and electrocatalytic properties of sonochemical synthesized PdAg nanoparticles, *Mater. Chem. Phys.*, **134**, 1013–1019.

74. Liu, H. and Manthiram, A. (2008). Tuning the electrocatalytic activity and durability of low cost $Pd_{70}Co_{30}$ nanoalloy for oxygen reduction reaction in fuel cells, *Electrochem. Commun.*, **10**, 740–744.

75. Choi, C. H., Park, S. H., and Woo, S. I. (2012). Oxygen reduction activity of Pd–Mn_3O_4 nanoparticles and performance enhancement by voltammetrically accelerated degradation, *Phys. Chem. Chem. Phys.*, **14**, 6842–6848.

76. Fernandez, J. L., Walsh, W. A., and Bard, A. J. (2005). Thermodynamic guidelines for the design of bimetallic catalysts for oxygen electroreduction and rapid screening by scanning electrochemical microscopy. M-Co (M: Pd, Ag, Au), *J. Am. Chem. Soc.*, **127**, 357–365.

77. Wang, Y. and Balbuena, P. B. (2005). Design of oxygen reduction bimetallic catalysts: ab-initio-derived thermodynamic guidelines, *J. Phys. Chem. B*, **109**, 18902–18906.

78. Wang, D., Xin, H. L., Yu, Y., Wang, H., Rus, E., Muller D. A., and Abruna, H. D. (2010). Pt-decorated PdCo@Pd/C core–shell nanoparticles

with enhanced stability and electrocatalytic activity for the oxygen reduction reaction, *J. Am. Chem. Soc.*, **132**, 17664–17666.

79. Rahul, R., Singh, R. K., and Neergat, M. (2014). Effect of oxidative heat-treatment on electrochemical properties and oxygen reduction reaction (ORR) activity of Pd-Co alloy catalysts, *J. Electroanal. Chem.*, **712**, 223–229.
80. Proshlyakov, D. A., Pressler, M. A., DeMaso, C., Leykam, J. F., DeWitt, D. L., and Babcock, G. T. (2000). Oxygen activation and reduction in respiration: involvement of redox-active tyrosine 244, *Science*, **24**, 1588–1591.
81. Dodelet, J. P. (2006). in N4-Macrocyclic Metal Complexes, Zagal, J. H., Bedioui, F., Dodelet, J. P. Eds. (Springer, New York), 83–147.
82. Morozan, A., Jousselme, B., and Palacin, S. (2011). Low-platinum and platinum-free catalysts for the oxygen reduction reaction at fuel cell cathodes, *Energy Environ. Sci.*, **4**, 1238–1254.
83. Alt, H., Binder, H., and Sandstede, G. (1973). Mechanism of the electrocatalytic reduction of oxygen on metal chelates, *J. Catal.*, **28**, 8–19.
84. Beck, F. (1977). The redox mechanism of the chelate-catalysed oxygen cathode, *J. Appl. Electrochem.*, **7**, 239–245.
85. Bogdanoff, P., Herrmann, I., Hilgendorff, M., Dorbandt, I., Fiechter, S., and Tributsch, H. (2004). Probing structural effects of pyrolysed CoTMPP-based electrocatalysts for oxygen reduction via new preparation strategies, *J. New Mater. Electrochem. Syst.*, **7**, pp 85–92.
86. Tylus, U., Jia, Q., Strickland, K., Ramaswamy, N., Serov, A., Atanassov, P., and Mukerjee, S. (2014). Elucidating oxygen reduction active sites in pyrolyzed metal–nitrogen coordinated non-precious-metal electrocatalyst systems, *J. Phys. Chem. C*, **118**, 8999–9008.
87. Gupta, S., Tryk, D., Bae, I., Aldred, W., and Yeager, E. (1989). Heat-treated polyacrylonitrile-based catalysts for oxygen electroreduction, *J. Appl. Electrochem.*, **19**, 19–27.
88. Lefevre, M., Dodelet, J. P., and Bertrand, P. (2000). Molecular oxygen reduction in PEM fuel cell conditions: ToF-SIMS analysis of Co-based electrocatalysts, *J. Phys. Chem. B*, **104**, 11238–11247.
89. Gasteiger, H. A., and Markovic, N. M. (2009). Just a dream—or future reality?, *Science*, **324**, 48–49.
90. Jaouen, F. and Dodelet, J. P. (2007). Average turn-over frequency of O_2 electro-reduction for Fe/N/C and Co/N/C catalysts in PEFCs, *Electrochim. Acta*, **52**, 5975–5984.

91. Jaouen, F., Proietti, E., Lefèvre, M., Chenitz, R., Dodelet, J.-P., Wu, G., Chung, H. T., Johnston, C. M., and Zelenay, P. (2011). Recent advances in non-precious metal catalysis for oxygen-reduction reaction in polymer electrolyte fuel cells, *Energy Environ. Sci.*, **4**, 114–130.

92. Kramm, U. I., Abs-Wurmbach, I., Herrmann-Geppert, I., Radnik, J., Fiechter, S., and Bogdanoff, P. (2011). Influence of the electron-density of FeN_4-centers towards the catalytic activity of pyrolyzed FeTMPPCl-based ORR-electrocatalysts, *J. Electrochem. Soc.*, **158**, B69–B78.

93. Yao, Z., Nie, H., Yang, Z., Zhou, X., Liu, Z., and Huang, S. (2012). Catalyst-free synthesis of iodine-doped graphene via a facile thermal annealing process and its use for electrocatalytic oxygen reduction in an alkaline medium, *Chem. Commun.*, **48**, 1027–1029.

94. Sheng, Z. H., Gao, H. L., Bao, W. J., Wang, F. B., and Xia, X. H. (2012). Synthesis of boron doped graphene for oxygen reduction reaction in fuel cells, *J. Mater. Chem.*, **22**, 390–395.

95. Yang, Z., Yao, Z., Li, G., Fang, G., Nie, H., Liu, Z., Zhou, X., Chen, X., and Huang, S. (2012). Sulfur-doped graphene as an efficient metal-free cathode catalyst for oxygen reduction, *ACS Nano*, **6**, 205–211.

96. Qu, L., Liu, Y., Baek, J.-B., and Dai, L. (2010). Nitrogen-doped graphene as efficient metal-free electrocatalyst for oxygen reduction in fuel cells, *ACS Nano*, **4**, 1321–1326.

97. Lai, L., Potts, J. R., Zhan, D., Wang, L., Poh, C. K., Tang, C., Gong, H., Shen, Z., Linc, J., and Ruoff, R. S. (2012). Exploration of the active center structure of nitrogen-doped graphene based catalysts for oxygen reduction reaction, *Energy Environ. Sci.*, **5**, 7936–7942.

98. Yang, S., Feng, X., Wang, X., and Müllen, K. (2011). Graphene-based carbon nitride nanosheets as efficient metal-free electrocatalysts for oxygen reduction reactions, *Angew. Chem. Int. Ed.*, **50**, 5339–5343.

99. Grger, B. N., Marković, N. M., and Ross, P. N. (1997). Temperature-dependent oxygen electrochemistry on platinum low-index single crystal surfaces in acid solutions, *Can. J. Chem.*,**75**, 1465–1471.

100. Qian, Y., Wen, W., Adcock, P. A., Jiang, Z., Hakim, N., Saha, M. S., and Mukerjee, S. (2008). PtM/C catalyst prepared using reverse micelle method for oxygen reduction reaction in PEM fuel cells, *J. Phys. Chem. C*, **112**, 1146–1157.

Chapter 7

Advances in Understanding the Effects on the Ethanol Electro-oxidation Reaction

Yixuan Wang
Department of Natural Science, Albany State University, Albany, Georgia 31705, USA
yixuan.wang@asurams.edu

This chapter primarily summarizes the recent advances in PtSn-based nanomaterials for direct alcohol fuel cells, such as PtSn-based trimetallic, oxides (TiO_2, SnO_2, and CeO_2)-doped bi- and trimetallic PtAu, and graphene/graphyne as supporting materials for the electrocatalysts of ethanol electro-oxidation reaction. The activity of the electrocatalyst PtSn could also vary with a few factors, like composition (Pt:Sn ratio), degree of alloying, roughness and size of particles, pH, as well as characteristics of electrolytes. These effects on the activity of the electrocatalyst and some open questions arising from experiments are also discussed.

Nanomaterials for Direct Alcohol Fuel Cell
Edited by Yixuan Wang

ISBN 978-981-4669-00-9 (Hardcover), 978-981-4669-01-6 (eBook)
www.panstanford.com

7.1 Introduction

A direct ethanol proton exchange membrane fuel cell (PEMFC) offers a promising alternative power source. It is superior to the most studied H_2-O_2 PEMFC in a few aspects. As a liquid fuel, it firstly avoids storage issue associated with H_2 fuel cells. A direct ethanol fuel cell also provides attractive mass energy density (8.1 kW·h/kg for ethanol vs. 0.42 kW·h/kg for H_2 1.5 wt% storage) [1]. In addition, ethanol is a renewable fuel since it can be produced in a great quantity from biomass, and it is much less toxic than methanol. However, the complete oxidation of ethanol to H_2O and CO_2 is much more difficult than those of H_2 as well as methanol because of the C–C bond in ethanol. Much effort has been put into searching for efficient electrocatalysts for the ethanol electro-oxidation reaction (EER) and considerable progress has also been made [1–11]. Among

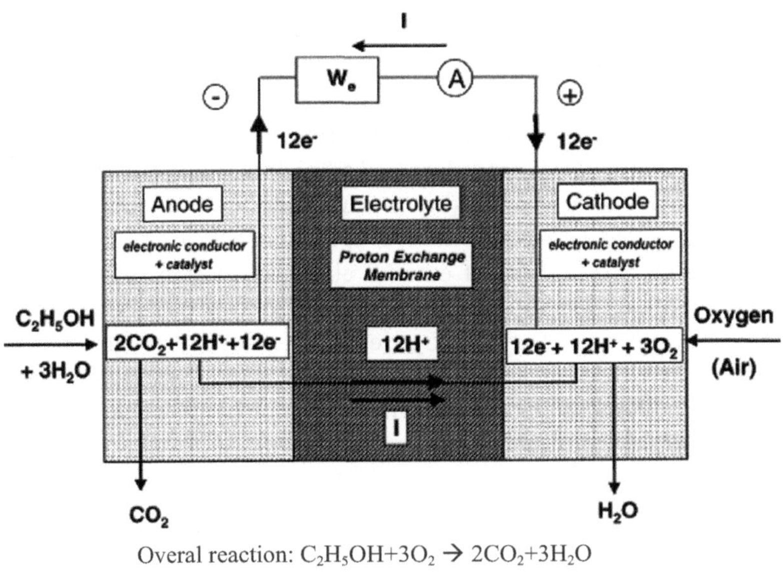

Figure 7.1 The working principle of a direct ethanol proton exchange membrane fuel cell in acidic medium. Reprinted with permission from Lamy et al. [12]. Copyright © 2002 Elsevier. In alkaline solution, the anode and cathode reactions are $C_2H_5OH+12OH^- \rightarrow 2CO_2+9H_2O+12e^-$ and $3O_2+6H_2O+12e^- \rightarrow 12OH^-$, respectively.

the pure transition metals, Pt had been recognized to be the most active catalyst for the EER in acidic solution; however, it is readily poisoned by adsorbed CO species and is hard to provoke C–C bond breaking for the desired complete oxidation of ethanol. The development of more efficient and less cost eletrocatalysts for the EER has been a challenge for wide application of the direct ethanol fuel cells.

In the acidic solutions the Pt-based bimetallic and trimetallic systems, and the addition of the oxides SnO_2 and CeO_2 to the Pt-based metallic systems are the most studied electrocatalysts for the EER [1, 4, 13–22]. In the recent years, it is also very interesting to find that Pt is more electroactive to the EER in alkaline media than in acidic solutions [8, 23], which stimulated investigations of the EER in alkaline media. The development of electrocatalysts for the EER and open questions arising from the existing investigations are summarized below.

7.2 PtSn, $PtSnO_2$, and Pt-Based Bi- and Trimetallic PtM_1M_2 (M_1 = Sn, Ru; M_2 = Ni, Rh, Bi, and Pd) in Acidic Media

The addition of Sn into Pt (deposited on graphite and other carbon materials, noted as PtSn/C thereafter) was widely recognized to enhance the electrocatalytic activity of Pt toward the EER with respect to increasing current density as well as to lowering the onset potential of ethanol oxidation by approximately 0.2 V compared to Pt alone [1, 4, 13, 14, 24]. Despite being less active than PtSn, a few other bimetallic nanoelectrocatalysts such as PtRu [17, 20, 25–27], PtNi [28], PtBi [20, 29], PdNi [30], PtAu [31, 32], PtAg [33], PtCu [34], etc., have also been demonstrated by a few reports to be more active than Pt to the EER.

Sn in the PtSn can be readily oxidized to SnO_2. Thus, Sn atoms and SnO_2 particles closely linked to Pt in PtSn/SnO_2 could offer a synergistic effect on the EER. The activity of the electrocatalyst PtSn could also vary with a few factors, like composition (Pt:Sn ratio), degree of alloying, roughness and size of particles, as well as pH and characteristics of electrolyte.

On the basis of Pt- or Pd-based bimetallic catalysts, a few trimetallic catalysts for the EER have also been established. The fuel cell tests have also demonstrated that the addition of Ni to Pt-Ru and Pt-Sn catalysts enhances the performance of the direct ethanol fuel cell, and for Pt-Ru-Ni catalysts the current density was 5 times greater than all other studied systems; however, Ni only duplicates the catalytic activity of the Pt-Sn catalyst [17]. The activities of a few carbon-supported ternary PtSn/RuM (M = Ni and Rh) electrocatalysts to the EER with a particle size of approximately 4 nm were recently compared [18], and it was found that PtRuNi showed a starting oxidation potential at a higher value than 0.64V. PtSnRh has slightly higher activity to the EER than PtSn. However, of PtSn/RuM (M = Ni, Rh) PtSnNi is the most active catalyst and provides the highest current density at potential of 0.5 V.

PtSnRh/C (50:40:10) and PtSn/C were shown by Spinace et al. to have a similar performance for ethanol oxidation at room temperature, while at 100° PtSnRh/C shows superior performance compared to PtSn/C and PtRh/C electrocatalysts [19]. PtRuBi/C (50:40:10) electrocatalyst prepared in two different ways shows superior performance for ethanol electro-oxidation at both room temperature and 100°C [20]. In situ mechanistic FTIR (in situ FTIR) analysis confirms an active 12-electron oxidation reaction channel to CO_2 at electrode potentials ≥450 mV/RHE, demonstrating the favorable efficiency of the PtRhSn Niggliite phase for C–C bond splitting [35]. The performance of EtOH electro-oxidation on Pd thin films as a catalyst layer in direct EtOH fuel cells (DEFCs) can be promoted by Co-electrodeposited Pd-Ni alloy thin films [36].

Nanostructured PtRuRh ternary catalysts towards the EER was also assessed by Datta et al. [21]. With higher catalytic property than Pt and the binary alloys, this ternary PtRhRu electrocatalyst could be a promising candidate for fuel cell applications. The proper introduction of Pd to the binary PtSn, the catalytic activity for EER decreases in the order PtSnPd (50:37.5:12.5) >PtSnPd (50:25:25) >PtSn (50:50), and the tin oxide phase has a promoting effect on the

EER and the incorporation of a low amount of Pd to Pt-Sn further enhances the activity for the EER [22].

A very recent interesting discovery is that the addition of nonmetal to the bimetallic may also enhance its activity to the EER. It was found that the electro-oxidation of ethanol on the Pd-Ni-P catalyst undergoes a 4-electron process to acetate [37]. The Pd-Ni-P catalyst contains the double number of electrocatalytically active sites (12.03%) compared with the Pd-Ni (6.04%) and Pd-black (5.12%) samples.

7.3 Effect of Alloying Degree and Size of PtSn on the Catalytic Behavior for the EER

Antolini et al. [15] recently reviewed the effects of PtSn structural characteristics, such as degree of alloying and Sn oxidation states, on the electrocatalytic activity for ethanol oxidation. Despite being still controversial in certain aspects, alloyed PtSn is generally believed to have superior effects to non-alloyed PtSn on the entire EER [15, 16]. For Pt-SnO_2 and alloyed $Pt_3Sn(111)$, Sn sites provides OH species to oxidize adsorbed residues from ethanol dehydrogenation, forming acetic acid (AA) and acetaldehyde (AAL); however, C–C and C–O bond breakings may be inhibited. The partially alloyed and Sn ad-atoms Pt_3Sn do not behave as well as the alloyed on the entire EER, but enhance C–C bond cleavage at lower potential than Pt. The behaviors are usually attributed to different bifunctional mechanism and electronic effects (ligand effects). However, Gupta et al. [38] recently claimed that alloying Pt with Sn lowers the activity of PtSn catalysts, possibly owing to the inhibition of ethanol adsorption/dehydrogenation on Pt. Therefore, to clarify the alloying effect, further studies at a molecular level need to be carried out.

It was showed that deposition of small SnO_2 nanoparticles onto the polycrystal-Pt electrode surface significantly enhances the catalytic performance of Pt [39]. IR spectroscopy measurements show the effect of the small SnO_2 particles on the removal of CO from the Pt surface, and suggest that the electrocatalytic C–C bond splitting activity of Pt, the key step in full conversion of ethanol to CO_2, is not affected by the addition of the –OH species provided by the SnO_2 particles. Also, controlled deposition of SnO_2 is needed to regulate the rate of partial oxidation to CH_3COOH compared to full oxidation to CO_2. The combined results provide new insight into the role of SnO_2 co-catalysts in promoting the electro-oxidation of ethanol to CO_2 on Pt surfaces.

7.4 Electro-oxidation of Ethanol on PtSn/C Doped by Metal Oxides (CeO_2, TiO_2) Electrocatalysts

The addition of CeO_2 to PtSn/C or $PtSnO_2$/C can generate rather active electrocatalysts to the EER, the ternary PtSn/CeO_2-C [7, 40, 41]. FTIR studies on the PtSn/C electrocatalyst showed that acetaldehyde (AAL) and acetic acid (AA) were the principal products, while on PtSn/CeO_2-C electrocatalyst the principal products were CO_2 and AA. It was speculated that the bifunctional mechanism is favored by the synergistic effect of SnO_2 and CeO_2 for generating oxides that promote the oxidation of AAL. The formation of CO_2 was explained by the C–C cleavage of the adsorbed AA. Nevertheless, the detailed mechanism about how CeO_2 cooperates with SnO_2 for the bifunctional mechanism and activates the C–C bond of adsorbed AA and AAL remains unknown.

The Pt/Ni-doped CeO_2/C catalyst was also found to exhibit remarkably high activity and stability towards the ethanol electro-oxidation in acidic media, which was attributed to higher oxygen-releasing capacity and stronger interaction of Ni-doped CeO_2 with Pt than pure CeO_2 nanoparticles [42]. The enhancement of the EER with Pd/C and TiO_2 was also addressed in alkaline media [43]. Nakagawa et al. reported the improved reaction kinetics and selectivity by the TiO_2-embedded carbon nanofiber support (PtRu-TiO_2/TECNF) for electro-oxidation of ethanol on PtRu nanoparticles [44]. Probably the design concept and synthetic strategy of metal-doped oxides used for fuel cell catalysts can be potentially extended to other catalytic fields.

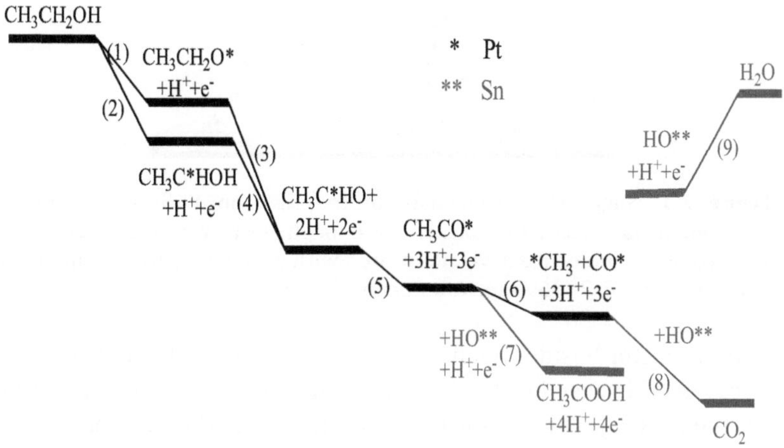

Figure 7.2 The preliminary mechanism for the EER on the PtSn/C, bifunctional mechanism.

7.5 Electrocatalytic Activity Enhancement Mechanism of PtSn to the EER in Acidic Solution

The mechanism of ethanol electro-oxidation on PtSn was widely interpreted with bifunctional mechanism and ligand effect [45, 46], where dissociative adsorption of ethanol occurs only on platinum sites and tin primarily promotes adsorption and dissociation of water to form OH-oxidizing intermediates from ethanol. Figure 7.2 shows the detailed mechanism proposed by Vigier et al. [2]. Other researchers suggested rather similar mechanism for the EOR [3, 4, 14, 47]. Although some interesting assumptions have been made, several essential issues remain uncertain. For example, even if H_2O dissociation can occur on Sn site at lower potential than on Pt alone. However, CH_3CO and CO, precursors of CO_2, were resulted from ethanol dissociation on Pt sites at high potential (0.6 V), which hardly explains why CO_2 can be produced at lower potential (0.3 V) on PtSn than on Pt alone. In addition, although Sn and Re have similar activity for water-dissociative adsorption, PtRe does not show itself to be as active as PtSn to the EOR. Thus, Sn may have

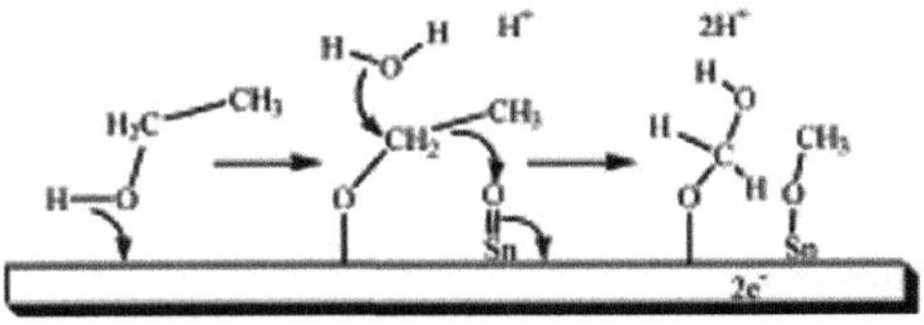

Figure 7.3 Suggested mechanism of the EER on Pt-SnO_2 composite. Whether surface accessible oxygen has enough electron affinity needs to be verified at the molecular level. Reprinted with permission from Mann et al. [5]. Copyright © 2006 American Chemical Society.

another factor besides ligand effect and enhancing H_2O dissociation because otherwise Re should display almost the same activity as Sn for the EER [2]. The significance of α- as well as β-dehydrogenation in ethanol electro-oxidation on Pt doped with Ru, Rh, Pd, Re, and others was recently discussed [48–50].

An alternative mechanism was suggested by Bocarsly et al. for the enhancement of PtSn to ethanol electro-oxidation [5]. They claimed that a tin-containing catalyst system would convert Sn to SnO_2 under the operational conditions of direct ethanol PEMFC, which supports Jiang and coworkers' finding that there obviously exists SnO_2 in PtSn catalysts [4]. Thus, the active catalyst in PtSn systems was considered to be Pt-SnO_2 composite. SnO_2 moiety on the composite surface may possess enough electron affinity to pull the methyl group off the α-carbon of ethanol, as shown in Fig. 7.3, during a nucleophilic attack by free water.

There is controversy on the mechanism of how Sn moiety enhances catalytic activity to the EOR in direct ethanol fuel cells. As briefly shown above, for the proposed electrocatalysts based on the PtSn (pH effect, trimetallic, PtSn/CeO_2-C, etc.), although bifunctional mechanism and ligand effect were widely applied to interpret their enhanced activity, there is still lack of convinced, unambiguous functional mechanism. In addition, although the EER was considerably improved by PtSn/C over Pt, the oxidation is still not complete and the majority of the oxidation products are C–C bond compounds such as AAL and AA. Partial oxidation greatly decreases the efficiency of a direct ethanol fuel cell. It is urgent to search for a new catalyst to break the C–C bond more readily—for example, adding a third metal to PtSn/C or PtRu to facilitate

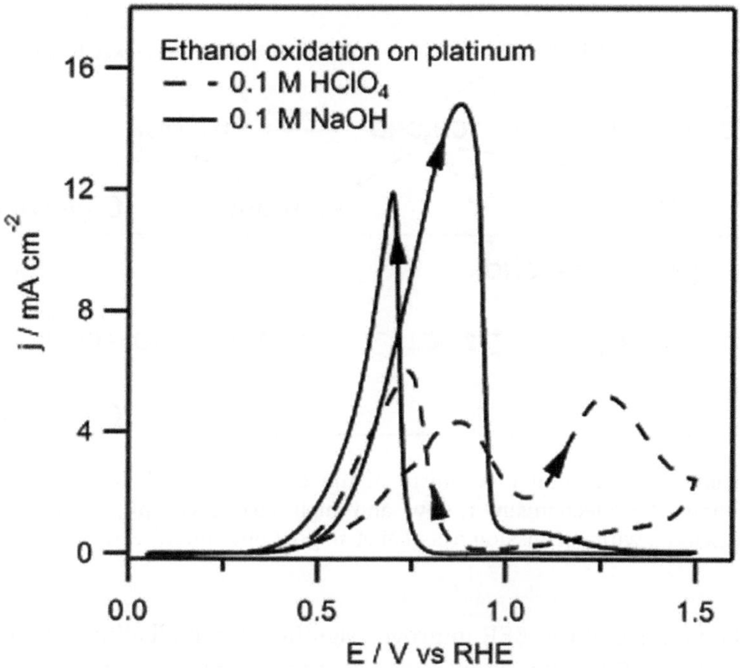

Figure 7.4 Cyclic voltammograms (first cycles) for the oxidation of 0.5 M ethanol in 0.1 M $HClO_4$ and 0.1 M NaOH on polycrystalline platinum. The voltammograms are recorded at 50 mV/s. The arrows indicate the scan direction. Reprinted with permission from Lai et al. [8]. Copyright © 2010 Elsevier.

the complete oxidation of ethanol. To advance the development of such catalysts, it is essential to gain a better understanding of the mechanism of EER on PtSn nanoparticles.

7.6 Effects of pH and Composition of Electrolyte on the Activity

Jiang et al. discovered that on both Pt/C and PtSn/C electrodes, the EER currents in the alkaline solutions were much higher than those in acidic solutions, and also that the onset potentials for the EER in alkaline solutions were less positive [8, 23], which shows that

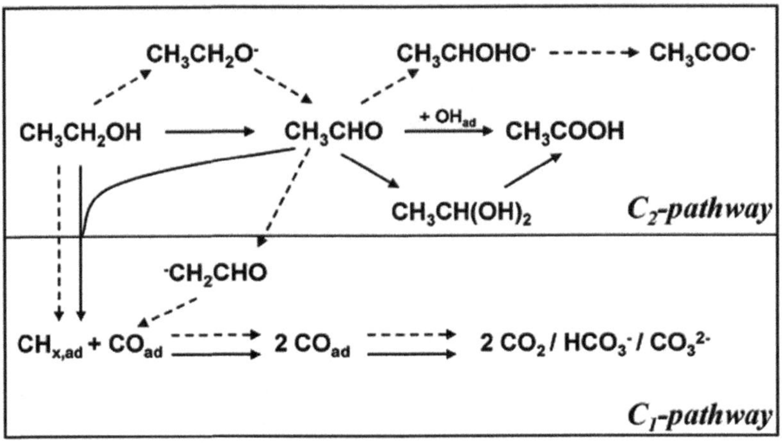

Figure 7.5 Reaction mechanism of the EER. Solid and dashed arrows denote the mechanism at low and high electrolyte pH, respectively. Reprinted with permission from Lai et al. [8]. Copyright © 2010 Elsevier.

the kinetics of the EER improves significantly in alkaline solutions [8, 9, 23]. Even though in acid solutions a significant improvement of PtSn/C over Pt/C was observed, the improvement was not so obvious in alkaline solutions [23].

The selectivity of the EER depends more strongly on the nature of electrolyte and, to a small extent, on pH [8]. Specifically, the C1 pathway (in Fig. 7.5), the broken C–C bond leading eventually to CO_2 is mainly noticeable on Pt electrode in the absence of strongly adsorbing anions, such as in $HClO_4$ or NaOH solutions. Lai et al. mainly attributed the higher activity of the electrocatalysts at high pH (>11) to the increased formation of acetaldehyde (AAL), which could be further oxidized via the germinal diol $CH_3CH(OH)_2^-$ at low pH and CH_3CHOHO^- at high pH as shown by the C2 pathway in Fig. 7.5. However, relatively little is known about whether acetaldehyde in the bulk solution or near the electrode is the major active species for its conversion to acetic acid. In addition, few reports analyze how pH of electrolyte affects adsorption/dehydrogenation processes along the EER. It is expected that density functional theory analysis, such as the highest occupied molecular orbital (HOMO), the lowest unoccupied molecular orbit

(LUMO) of the catalysts (Pt and PtSn), and their interactions, would greatly help understand the pH effect.

Pd appears to be a better catalyst for the selective oxidation of alcohols in alkaline media, but where poisoning by the adsorbed CO plays a significant role, such as in methanol oxidation, Pt is still the preferred catalyst [10]. Alkaline Pd-Sn/C displayed better electrocatalytic activity and stability towards poisoning than Pd-Ru-Sn/C and Pt-Ru/C nanocatalysts [51]. Silver can also promote the palladium catalysis for ethanol electro-oxidation in an alkaline electrolyte [33].

7.7 Unexpected Activity of Gold (Au) to the EER in Alkaline Media

Although Au has rarely been recognized as an electrocatalyst for the EER in acidic solutions due to its poor ability to break C–C bond, most recently the electrocatalytic oxidation of alcohols on gold in alkaline media was proven to be amazing [52]. It was discovered that the first dehydrogenation of ethanol is base-catalyzed, and the second deprotonation is fast but gold-catalyzed. The high electrocatalytic activity to the EER on Au compared to Pt is attributed to the high resistance of gold toward the formation of poisoning surface oxides. These results indicate that base catalysis is the main driver behind the high oxidation activity in alkaline media, and not the catalyst interaction with hydroxide [52].

Electrochemical and fuel cell evaluation of PtAu/C electrocatalysts for ethanol electro-oxidation were discussed in alkaline media [31]. Electrochemical characterization of the PtAu/C materials suggested the PtAu/C (50:50) as the most promising material for EtOH electro-oxidation, while experiments in single fuel cell suggested PtAu/C (70:30). The discrepancy can be explained by the different electrodes since PtAu/C (50:50) yields a much thicker electrode than PtAu/C (70:30).

PtAu catalysts supported with different nanostructures, such as core shell structured (Au@Pt/C) and alloy nanoparticles (AuPt/C), shows different activity to the EER [32]. The stronger Pt-Au

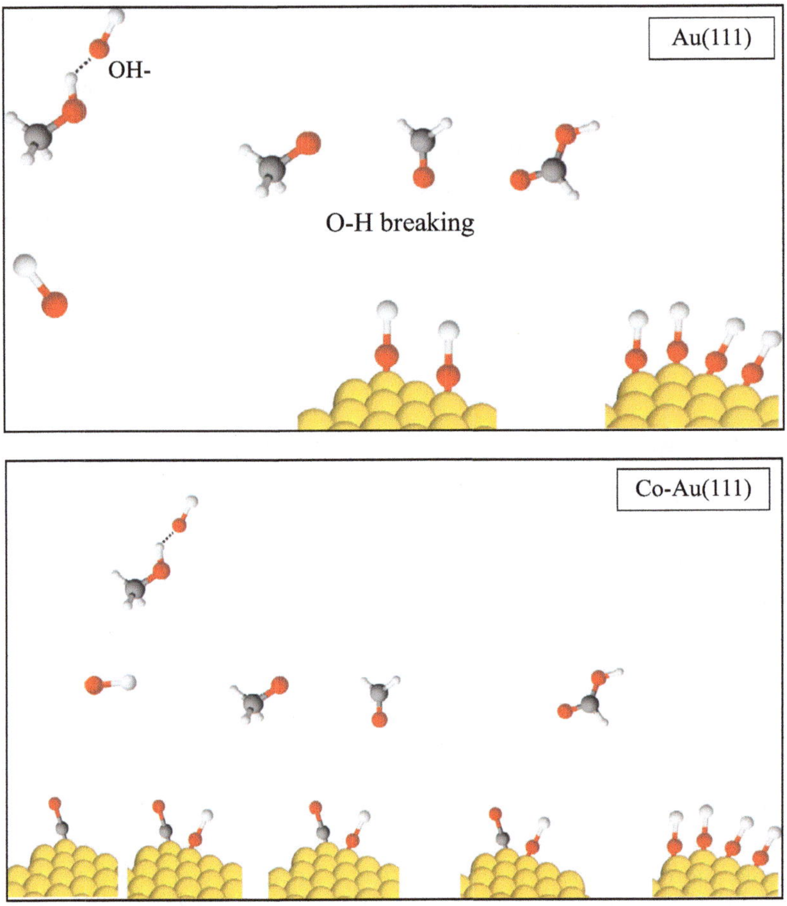

Figure 7.6 Illustration of the mechanism of alcohol oxidation on Au(111) as catalyzed by CO_{ads}. CO_{ads} *lowers the potential at which OH adsorbs* on Au(111) and thereby allows for the oxidation of methoxy to formaldehyde at a significantly lower potential. Red atoms represent oxygen, white atoms hydrogen, grey atoms carbon, and gold atoms gold. Reprinted by permission from Rodriguez et al. [11]. Copyright © 2012 Macmillan Publishers Ltd.

interaction and the expanded Pt lattice parameter in alloyed PtAu/C contributed to the higher specific activity towards EtOH electro-oxidation of PtAu/C compared to Au@Pt/C.

More interestingly, for heterogeneous catalysis and electrocatalysis, adsorbed CO typically acts as a poison or poisoning intermediate in the oxidation of alcohols. However, it was reported recently that the CO adsorbed on a Au(111) surface in alkaline media can act as a promoter for the electrocatalytic oxidation of alcohols [11]. In comparison with bare Au(111), the onset potential for methanol oxidation is significantly lower in the presence of adsorbed CO, and formation of the main methanol oxidation products—formaldehyde and formic acid—is enhanced, which due to the presence of adsorbed CO promotes beta-hydrogen elimination, i.e., C–H bond breaking [11].

7.8 Extremely High Electrocatalytic Activity of Pt/Au Structure to the EER in Alkaline Media

Pt/Au and Pd/Au surface interfaces show very high activity in electrocatalytic oxidation of alcohols in alkaline media [53]. The peak current density of ethanol electro-oxidation on the Au_3Pt_1 nanodendrite-modified glassy carbon electrode (Au_3Pt_1 electrode) is approximately 16, 12.5, and 4.5 times higher than those on thc polycrystalline Pt electrode, polycrystalline Au electrode, and Au nanodendrite-modified glassy carbon electrode (Au dendrites electrode), respectively [54]. The oxidation peak potential of ethanol electro-oxidation on the Au_3Pt_1 electrode is about 299 and 276 mV lower than those on the polycrystalline Au electrode and Au dendrites electrode, respectively. The Au/Pt bimetallic nanodendrites may find potential application in alkaline direct ethanol fuel cells [54]. Cherevko et al. attributed the extreme electroactivity of the Pt/Au structures toward the EOR to the active surface states, i.e., undercoordinated interface atoms (Fig. 7.7) [53] and proposed a few open questions: Why is the Pt/Au system so active in alkaline solution and totally inactive in acidic solution? What is the nature of the active states atoms and how does the coordination number of an atom change its chemical properties?

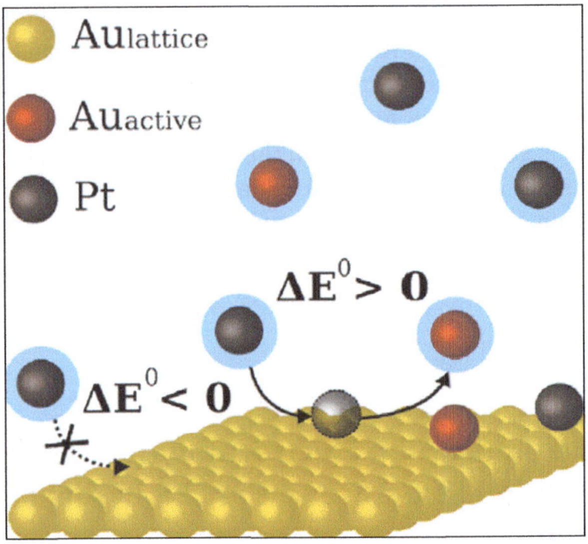

Figure 7.7 Schematic of the proposed mechanism of the galvanic displacement deposition of Pt or Pd on the surface of the gold electrode. Reprinted with permission from Cherevko et al. [53]. Copyright © 2012 Elsevier.

7.9 Graphene, Defective Graphene, and Other Novel Nanomaterials for the EER

Graphene, a new allotrope of carbon, is one planar sheet of sp^2-bonded carbon atoms arranged in a hexagonal lattice. Like carbon nanotubes, graphene sheets have demonstrated a variety of applications, including high electron mobility, and the enhancement of mechanical and electrical properties of composite materials. During production of graphene from the thermal expansion of graphite oxide (GO), some carbon atoms are missing to form defective graphene [55]. Because of carbon vacancies the defective graphene may demonstrate significant influences on the chemical and physical characteristics of graphene, for example, stronger adsorption on the defect sites [56]. A previous experimental study also suggested that the presence of defects on graphene substrates allows for controlled morphology of iron nanoparticles [57]. The most common defects of graphene include monovacancies, multivacancies, pentagon heptagon pairs, and adatoms [58]. According to experimental and theoretical results [59, 60], the double vacancy 5-8-5 defect (two pentagonal rings and one octagonal ring) is considered to be the most stable defect in the graphenes where the functional groups are the hydroxyls and epoxies. Very recently, the adsorptions of metal clusters on the monovacancy and double-vacancy defective (5-8-5 defect) graphenes were investigated [61–64]. The structural and electronic properties of the nanoparticles adsorbed on the defective graphene usually show peculiarities. The defective sites sever as anchoring points for the nanoparticles, and undercoordinated neighboring carbons further strengthen the binding of the nanoparticles to graphene layer. Catalytic reactivity of the adsorbed nanoparticles may be also enhanced. Lim found that the most stable conformation of Pt_{13} on monovacancy defective graphene has D_{4h} symmetry, rather than the isolated I_h symmetry, and the Pt_{13} donates electron to the defective graphene and the adsorbed O_2 [62]. To the best of our knowledge, the adsorption of Pt-based bimetallic on defective graphene and their activity to the ethanol oxidation have not been well addressed in spite of a few reports, as described below.

In terms of the electrocatalytic properties, the electrochemical activity of Pt-Sn/graphene for the EtOH oxidation reaction was improved with increasing Sn content due to the bifunctional mechanism and ligand effects [65]. A series of graphene (G)-supported Ni_xPd_{100-x} binary alloyed catalysts (BACs) were prepared with variation of Ni and Pd metal loading through a facile chemical reduction method [66]. In 1 M KOH electrolyte the higher catalytic activity for EER was increased in the order $Ni_{75}Pd_{25}/G > Ni_0Pd_{100}/G > Ni_{25}Pd_{75}/G > Ni_{50}Pd_{50}/G$. The overall parameters of EOR study determined that the $Ni_{50}Pd_{50}/G$ was more favorable in various ethanol (EtOH) concentrations and scan rates [66].

The Pt-Cu bimetallic nanoparticles supported on reduced graphene oxide (Pt-Cu/RGO) was characterized with UV-vis spectrophotometer, X-ray diffraction, transmission electron microscopy, and XPS and its catalytic behavior for the direct oxidation of methanol was investigated [34]. Compared to Pt/RGO and Pt/C catalysts, Pt-Cu/RGO hybrids exhibited markedly superior catalytic activity for the electrocatalytic oxidation of methanol and ethanol.

This improved catalytic activity can be attributed to the dendritic structure of the Pt-Cu bimetallic nanoparticles [34].

Moreover, although non-precious metal electrocatalysts for the oxygen reduction reaction of PEM fuel cells have been widely explored [67], there are few reports on such sort of electrocatalysts for the EER. It is also known that to enhance the efficiency, the electrocatalysts are usually deposited on the C-based substrates (PtSn/C), which significantly affects the size of the nanoparticles. Two-dimesional graphene is a promising new material because of high-speed movement of electrons in graphene. Basically, boron doping in graphene introduces hole; that is, it is hole doping. Compared with the pure armchair graphene nanoribbon, the Fermi level is shifted down in the B-doped armchair graphene nanoribbon and the calculated band structure shows a metallic behavior [68]. We are also eager to get an insight into the behavior of B-, N-, O-, and F-doped graphene to the EER.

N-doped C spheres (NCSs) were prepared as a new catalyst support for EtOH electro-oxidation through carbonization of poly(o-phenylenediamine) spheres. The NCS-supported Pd nanocatalyst exhibited higher current density and better stable life for EtOH electro-oxidation in alkaline media than Pd catalysts supported on activated C- and N-doped C nanotubes. The catalytic current of the Pd/NCS composite was even comparable to commercial Pt catalyst (40% Pt loading) [69].

Very recently, another allotrope of C, graphyne (GY), was also investigated for the catalysts of the DEFC, mostly for oxygen reduction reaction. Similar to graphene, GY is also a two-dimensional carbon network, but it replaces some =C=C= bonds in graphene by acetylene linkages (−C≡C−). Because of the −C≡C− insertion, some carbon atoms in graphyne are positively charged to facilitate O_2 dissociation. It was also found that the first two-hydrogen addition produces a water molecule, and then the next two hydrogens furnish another water formation accompanied by the recovery of graphyne planarity. The ORR process on three graphyne models exhibits an efficient four-electron process [70, 71]. The transition metal decorated graphyne was also proposed to be an efficient catalyst for oxygen reduction reaction [72]. Iron and cobalt decorated graphyne are found to be more reactive toward oxygen adsorption, whereas the nickel-decorated system is inactive. In acid medium, both iron- and cobalt-decorated systems were observed to follow a more efficient four-electron reduction path. However, in alkaline medium, the reduction on an iron-decorated system is through a four-electron reduction path, while in the cobalt system it is through a two-electron reduction path with the formation of a hydroperoxide anion. The free-energy profiles also show that these systems are more efficient as catalysts for oxygen reduction in alkaline medium as compared to acid medium.

References

1. Vigier, F., Coutanceau, C., Perrard, A., Belgsir, E. M., Lamy, C. (2004). Development of anode catalysts for a direct ethanol fuel cell. *J. Appl. Electrochem.*, **34**, 439–446.
2. Vigier, F., Coutanceau, C., Hahn, F., Belgsir, E. M., Lamy, C. (2004). On the mechanism of ethanol electro-oxidation on Pt and PtSn catalysts: electrochemical and in situ IR reflectance spectroscopy studies. *J. Electroanal. Chem.*, **563**, 81–89.
3. Song, S., Tsiakaras, P. (2006). Recent progress in direct ethanol proton exchange membrane fuel cells (DE-PEMFCs). *Appl. Catal. B: Environ.*, **63**, 187–193.
4. Jiang, L., Sun, G., Sun, S., Liu, J., Tang, S., Li, H., Zhou, B., Xin, Q. (2005). Structure and chemical composition of supported Pt-Sn electrocatalysts for ethanol oxidation. *Electrchim. Acta*, **50**, 5384–5389.
5. Mann, J., Yao, N., Bocarsly, A. B. (2006). Characterization and analysis of new catalysts for a direct ethanol fuel cell. *Langmuir*, **22**, 10432–10436.
6. Rousseau, S., Coutanceau, C., Lamy, C., Leger, J.-M. (2006). Direct ethanol fuel cell (DEFC): electric performances and reaction products distribution under operating conditions with different platinum-based anodes. *J. Power Sources*, **158**, 18–24.
7. Neto, O. A., Farias, L. A., Dias, R. R., Brandalise, M., Linardi, M., Spinace, E. V. (2008). Enhanced electro-oxidation of ethanol using PtSn/CeO2-C electrocatalyst prepared by an alcohol-reduction process. *Electrochem. Commun.*, **10**, 1315.
8. Lai, S. C. S., Kleijn, S. E. F., Öztürk, F. T. Z., Vellinga, V. C. V. R., Koning, J., Rodriguez, P., Koper, M. T. M. (2010). Effects of electrolyte pH and

composition on the ethanol electro-oxidation reaction. *Catal. Today*, **154**, 92–104.

9. Kutz, R. B., Braunschweig, B., Mukherjee, P., Dlott, D. D., Wieckowski, A. (2011). Study of ethanol electrooxidation in alkaline electrolytes with isotope labels and sum-frequency generation. *J. Phys. Chem. Lett.*, **2**, 2236–2240.
10. Santasalo-Aarnio, A., Kwon, Y., Ahlberg, E., Kontturi, K., Kallio, T., Koper, M. T. M. (2011). Comparison of methanol, ethanol and iso-propanol oxidation on Pt and Pd electrodes in alkaline media studied by HPLC. *Electrochem. Commun.*, **13**, 466–469.
11. Rodriguez, P., Kwon, Y., Koper, M. T. M. (2012). The promoting effect of adsorbed carbon monoxide on the oxidation of alcohols on a gold catalyst. *Nat. Chem.*, **4**, 177–182.
12. Lamy, C., Lima, A., LeRhun, V., Delime, F. (2002). Recent advances in the development of direct alcohol fuel cells (DAFC). *J. Power Sources*, **105**, 283–296.
13. Lamy, C., Rousseau, S., Belgsir, E. M., Coutanceau, C., Leger, J.-M. (2004). Recent progress in the direct ethanol fuel cell: development of new platinum-tin electrocatalysts. *Electrchim. Acta*, **49**, 3901–3908.
14. Song, S. Q., Zhou, W. J., Zhou, Z. H., Jiang, L. H., Sun, G. Q., Xin, Q., Leontidis, V., Kontou, S., Tsiakaras, P. (2005). Direct etahnol PEM fuel cells: the case of platinum based anods. *Int. J. Hydrogen Engergy*, **30**, 995–1001.
15. Antolini, E., Gonzalez, E. R. (2011). Effect of synthesis method and structural characteristics of Pt-Sn fuel cell catalysts on the electro-oxidation of CH3OH and CH3CH2OH in acid medium. *Catal. Today*, **160**, 28–38.
16. Zhu, M., Sun, G., Xin, Q. (2009). Effect of alloying degree in PtSn catalyst on the catalytic behavior for ethanol electro-oxidation. *Electrochim. Acta*, **54**, 1511.
17. Ribadeneira, R. E. H., B. A. (2010). Co-catalytic effect of nickel in Pt-Ru/C and Pt-Sn/C electrocatalysts for ethanol electrooxidation. *Lat. Am. Appl. Res.*, **40**, 113.
18. Bonesi, A., Garaventa, G., Triaca, W. E., Castro Luna, A. M. (2008). Synthesis and characterization of new electrocatalysts for ethanol oxidation. *Int. J. Hydrogen Energy*, **33**, 3499–3501.
19. Spinace, E. V., Dias, R. R., Brandalise, M., Linardi, M., Oliveira Neto, A. (2010). Electro-oxidation of ethanol using PtSnRh/C electrocatalysts prepared by an alcohol-reduction process. *Ionics*, **16**, 91.

20. Brandalise, M., Tusi, M. M., Piasentin, R. M., Linardi, M., Spinace, E. V., Oliveira Neto, A. (2010). PtRu/C and PtRuBi/C electrocatalysts prepared in two different ways by borohydride reduction for ethanol electro-oxidation. *Int. J. Electrochem. Sci.*, **5**, 39–45.
21. Datta, J., Singh, S. (2009). In *Central Regional Meeting of the American Chemical Society*. Cleveland, OH, United States, CRM-130.
22. Lee, E., Park, I.-S., Manthiram, A. (2010). Synthesis and characterization of Pt-Sn-Pd/C catalysts for ethanol electro-oxidation reaction. *J. Phys. Chem. C*, **114**, 10634.
23. Jiang, L. H., A., Chu, D., Chen, R. (2010). Ethanol electro-oxidation on Pt/C and PtSn/C catalysts in alkaline and acid solutions. *Int. J. Hydrogen Energy*, **35**, 365–372.
24. Herranz, T. I., Maria, Gomez de la Fuente, Jose L., Perez-Alonso, Francisco J., Pena, Miguel A., Cabot, Andreu, Rojas, Sergio (2014). In situ study of ethanol electrooxidation on monodispersed Pt3Sn nanoparticles. *ChemElectroChem*, **1**, 885–895.
25. Xu, W., Lu, T., Liu, C., Xing, W. (2006). Electrooxidation of COad intermediated from methanol oxidation on polycrystalline Pt electrode. *J. Phys. Chem. B*, **110**, 4802–4807.
26. Basnayake, R., Li, Z., Katar, S., Zhou, W., Rivera, H., Smotkin, E. S., Casadonte, J., Korzeniewski, C. (2006). PtRu nanoparticle electrocatalyst with bulk alloy properties prepared through a sonochemical method. *Langmuir*, **22**, 10446–10450.
27. Chatterjee, M., Chatterjee, A., Ghosh, S., Basumallick, I. (2009). Electro-oxidation of ethanol and ethylene glycol on carbon-supported nano-Pt and -PtRu catalyst in acid solution. *Electrochim. Acta*, **54**, 7299–7304.
28. Park, K.-W., Choi, J.-H., Kwon, B.-K., Lee, S.-A., Sung, Y.-E., Ha, H.-Y., Hong, S.-A., Kim, H., Wieckowski, A. (2002). Chemical and electronic effects of Ni in Pt/Ni and Pt/Ru/Ni alloy nanoparticles in methanol electrooxidation. *J. Phys. Chem. B*, **102**, 1869–1877.
29. Casado-Rivera, E., Volpe, D. J., Alden, L., Lind, C., Angelo, A. C. D., DiSalvo, F. J., Abruna, H. D. (2004). Electrocatalytic activity of ordered intermetallic phases for fuel cell applications. *J. Am. Chem. Soc.*, **126**, 4043–4049.
30. Awasthi, R., Singh, R. N. (2010). Synthesis and characterization of nano structured Pd-Ni and Pd-Ni-C composites towards electrooxidation of alcohols. *Open Catal. J.*, **3**, 70.
31. da Silva, S. G. S., Julio Cesar M., Buzzo, Guilherme S., De Souza, Rodrigo F. B., Spinace, Estevam V., Neto, Almir O., Assumpcao, Monica

H. M. T. (2014). Electrochemical and fuel cell evaluation of PtAu/C electrocatalysts for ethanol electro-oxidation in alkaline media. *Int. J. Hydrogen Energy*, **39**, 10121–10127.

32. Zhou, W. L., Miao, Zhang, Lan, Chan, Siew Hwa (2014). Supported PtAu catalysts with different nano-structures for ethanol electrooxidation. *Electrochim. Acta*, **123**, 233–239.
33. Feng, Y.-Y. L., Zeng-Hua, Kong, Wei-Qing, Yin, Qian-Ying, Du, Li-Xia (2014). Promotion of palladium catalysis by silver for ethanol electro-oxidation in alkaline electrolyte. *Int. J. Hydrogen Energy*, **39**, 2497–2504.
34. Li, F. G., Yongqin, Chen, Mingxi, Qiu, Haixia, Sun, Xiying, Wang, Wei, Liu, Yu, Gao, Jianping (2013). Comparison study of electrocatalytic activity of reduced graphene oxide supported Pt-Cu bimetallic or Pt nanoparticles for the electrooxidation of methanol and ethanol. *Int. J. Hydrogen Energy*, **38**, 14242–14249.
35. Erini, N. L., Rameshwori, Petkov, Valeri, Baranova, Elena A., Yang, Ruizhi, Teschner, Detre, Huang, Yunhui, Brankovic, Stanko R., Strasser, Peter (2014). Ethanol electro-oxidation on ternary platinum-rhodium-tin nanocatalysts: insights in the atomic 3D structure of the active catalytic phase. *ACS Catal.*, **4**, 1859–1867.
36. Chang, C.-M. H., Mu-Tao, Kang, Wei-Chieh, Whang, Thou-Jen (2014). Study of the Co-electrodeposited Pd-Ni alloy thin film and its performance for ethanol electro-oxidation. *J. Electrochem Soc.*, **161**, D552–D557.
37. Jiang, R. T., Dat T., McClure, Joshua P., Chu, Deryn (2014). A class of (Pd-Ni-P) electrocatalysts for the ethanol oxidation reaction in alkaline media. *ACS Catal.*, **4**, 2577–2586.
38. Gupta, S. S., Singh, S., Datta, J. (2009). Promoting role of unalloyed Sn in PtSn binary catalysts for ethanol electro-oxidation. *Mater. Chem. Phys.*, **116**, 223–228.
39. Magee, J. W. Z., Wei-Ping, White, Michael G. (2014). Promotion of Pt surfaces for ethanol electro-oxidation by the addition of small SnO2 nanoparticles: activity and mechanism. *Appl. Catal. B: Environ.*, **152–153**, 397-4-2.
40. Neto, O. A., Linardi, M., Anjos, D. M., Tremiliosi-Filho, G., Spinace, E. V. (2009). Electro-oxidation of ethanol on PtSn/CeO2-C electrocatalyst. *J. Appl. Electrochem.*, **39**, 1153–1156.
41. He, Q., Mukerjee, S., Shyam, B., Ramaker, D., Parres-Esclapez, S., Illan-Gomez, M. J., Bueno-Lopez, A. (2009). Promoting effect of CeO2 in

the electrocatalytic activity of rhodium for ethanol electro-oxidation. *J. Power Sources*, **193**, 408–415.

42. Tan, Q. D., Chunyu, Sun, Yongrong, Du, Lei, Yin, Geping, Gao, Yunzhi (2014). Nickel-doped ceria nanoparticles for promoting catalytic activity of Pt/C for ethanol electrooxidation. *J. Power Sources*, **263**, 310–314.
43. Silva, J. C. M., Buzzo, G. S., De Souza, R. F. B., Spinace, E. V., Neto, A. O., Assumpcao, M. H. M. T. (2015). Enhanced electrooxidation of ethanol using Pd/C + TiO2 electrocatalysts in alkaline media. *Electrocatalysis*, **6**, 86–91.
44. Nakagawa, N. I., Yudai, Tsujiguchi, Takuya, Ishitobi, Hirokazu (2014). Improved reaction kinetics and selectivity by the TiO2-embedded carbon nanofiber support for electro-oxidation of ethanol on PtRu nanoparticles. *J. Power Sources*, **248**, 330–336.
45. Watanabe, M., Motoo, S. (1975). Electrocatalysis by ad-atoms. *J. Electroanal. Chem.*, **60**, 275–283.
46. Iwasita, T., Pastor, E. (1994). A DEMS and FTIR spectroscopic investigation of adsorbed ethanol on polycrystalline platinum. *Electrochim. Acta*, **39**, 531.
47. Lux, K. W. C., Elton J. (2006). Lanthanide-platinum intermetallic compounds as anode electrocatalysts for direct ethanol PEM fuel cells. II. Performance of LnPt2 (Ln = Ce, Pr) nanopowders in an operating PEM fuel cell. *J. Electrochem. Soc.*, **153**, A1139–A1147.
48. Sheng, T. L., Wen-Feng, Hardacre, Christopher, Hu, P. (2014). Significance of β-dehydrogenation in ethanol electro-oxidation on platinum doped with Ru, Rh, Pd, Os and Ir element. *Phys. Chem. Chem. Phys.*, **16**, 13248–13254.
49. Xu, Z., Wang, Y. W. (2013) *Applications of Molecular Modeling to Challenges in Clean Energy, ACS Symposium Series* Fitzgerald, G., Govind, N., Chapter 8 "Understanding electrocatalytic activity enhancement of bimetallic particles to ethanol electro-oxidation: ethanol adsorption and decomposition on PtnM (n=6 and 9, M=Pt, Ru, and Sn)" (American Chemical Society), **1113**, 135–151.
50. Xu, Z.-F., Wang, Y. (2011). Effects of alloyed metal on the catalysis activity of Pt for ethanol partial oxidation: adsorption and dehydrogenation on Pt3M (M = Pt, Ru, Sn, Re, Rh, and Pd). *J. Phys. Chem. C*, **115**, 20565–20571.
51. Modibedi, R. M., Masombuka, T., Mathe, M. K. (2011). Carbon supported Pd-Sn and Pd-Ru-Sn nanocatalysts for ethanol electro-oxidation in alkaline medium. *Int. J. Hydrogen Energy*, **36**, 4664–4672.

52. Kwon, Y., Lai, S. C. S., Rodriguez, P., Koper, M. T. M. (2011). Electrocatalytic oxidation of alcohols on gold in alkaline media: base or gold catalysis. *J. Am. Chem. Soc.*, **133**, 6914–6917.
53. Cherevko, S., Kulyk, N., Chung, C.-H. (2012). Utilization of surface active sites on gold in preparation of highly reactive interfaces for alcohols electrooxidation in alkaline media. *Electrochim. Acta*, **69**, 190–196.
54. Han, X., Wang, D., Liu, D., Huang, J., You, T. (2012). Synthesis and electrocatalytic activity of Au/Pt bimetallic nanodendrites for ethanol oxidation in alkaline medium. *J. Colloid Interface Sci.*, **367**, 342–347.
55. Kudin, K. N., Ozbas, B., Schniepp, H. C., Prud'homme, R. K., Aksay, I. A., Car, R. (2008). Raman spectra of graphite oxide and functionalized graphene sheets. *Nano Lett.*, **8**, 36–41.
56. Boukhvalov, D. W., Katsnelson, M. I. (2008). Chemical functionalization of graphene with defects. *Nano Lett.*, **8**, 4373–4379.
57. Wang, H. L., Robinson, J. T., Diankov, G., Dai, H. J. (2010). Nanocrystal growth on graphene with various degrees of oxidation. *J. Am. Chem. Soc.*, **132**, 3270–3271.
58. Hashimoto, A., Suenaga, K., Gloter, A., Urita, K., Iijima, S. (2004). Direct evidence for atomic defects in graphene layers. *Nature*, **430**, 870–873.
59. Schniepp, H. C., Li, J. L., McAllister, M. J., Sai, H., Herrera-Alonso, M., Adamson, D. H., Prud'homme, R. K., Car, R., Saville, D. A., Aksay, I. A. (2006). Functionalized single graphene sheets derived from splitting graphite oxide. *J. Phys. Chem. B*, **110**, 8535–8539.
60. Lee, G.-D., Wang, C. Z., Yoon, E., Hwang, N.-M., Kim, D.-Y., Ho, K. M. (2005). Diffusion, coalescence, and reconstruction of vacancy defects in graphene layers. *Phys. Rev. Lett.*, **95**, 205501.
61. Lim, D.-H., Negreira, A. S., Wilcox, J. (2011). DFT studies on the interaction of defective graphene-supported Fe and Al nanoparticles. *J. Phys. Chem. C*, **115**, 8961–8970.
62. Lim, D.-H.,Wilcox, J. (2011). DFT-based study on oxygen adsorption on defective graphene-supported Pt nanoparticles. *J. Phys. Chem. C*, **115**, 22742–22747.
63. Kou, R., Shao, Y., Mei, D., Nie, Z., Wang, D., Chongmin Wang, Viswanathan, V. V., Park, S., Aksay, I. A., Lin, Y., Wang, Y., Liu, J. (2011). Stabilization of electrocatalytic metal nanoparticles at metal-metal oxide-graphene triple junction points. *J. Am. Chem. Soc.*, **133**, 2541–2547.
64. Kim, G., Jhi, S.-H. (2011). Carbon monoxide-tolerant platinum nanoparticle catalysts on defect-engineered graphene. *ACS Nano*, **5**, 805–810.

65. Wang, Y., Wu, G., Wang, Y., Wang, X. (2014). Effect of water content on the ethanol electro-oxidation activity of Pt-Sn/graphene catalysts prepared by the polyalcohol method. *Electrochim. Acta*, **130**, 135–140.

66. Ahmed, M. S. J., Seungwon (2014). Highly active graphene-supported NixPd100-x binary alloyed catalysts for electro-oxidation of ethanol in an alkaline media. *ACS Catal.*, **4**, 1830–1837.

67. Liu, G., Li, X., Ganesan, P., Popov, B. N. (2009). Development of non-precious metal oxygen-reduction catalysts for PEM fuel cells based on N-doped ordered porous carbon. *Appl. Catal. B: Environ.*, **93**, 156–165.

68. Martins, T. B., Miwa, R. H., Silva, A. J. R. d., Fazzio, A. (2007). Electronic and transport properties of boron-doped graphene nanoribbons. *Phys. Rev. Lett.*, **98**, 196803.

69. Li, Y. Y., Li, Zhang, Lin-Qun, Liu, An-Ran, Zhang, Yuan-Jian, Liu, Song-Qin (2014). Palladium nanoparticles supported on nitrogen-doped carbon spheres as enhanced catalyst for ethanol electro-oxidation. *J. Electroanal. Chem.*, **730**, 65–68.

70. Kang, B., Lee, J. Y. (2014). Graphynes as promising cathode material of fuel cell: improvement of oxygen reduction efficiency. *J. Phys. Chem. C*, **118**, 12035–12040.

71. Yong, K. B. L. H. L. J. (2014). Oxygen adsorption on single layer graphyne: a DFT study. *Phys. Chem. Chem. Phys.*, **16**, 974–980.

72. Srinivasu, K., Ghosh, S. K. (2013). Transition metal decorated graphyne: an efficient catalyst for oxygen reduction reaction. *J. Phys. Chem. C*, **117**, 26021–26028.

Index

An environmentally friendly book printed and bound in England by www.printondemand-worldwide.com

PEFC Certified

This product is from sustainably managed forests and controlled sources

www.pefc.org

PEFC/16-33-415

This book is made of chain-of-custody materials; FSC materials for the cover and PEFC materials for the text pages.